DESCRIPTIVE
MICROMETEOROLOGY

ADVANCES IN GEOPHYSICS

EDITED BY

H. E. LANDSBERG

U.S. ENVIRONMENTAL SCIENCE SERVICES
ADMINISTRATION
WASHINGTON, D.C.

J. VAN MIEGHEM

ROYAL BELGIAN METEOROLOGICAL
INSTITUTE
UCCLE, BELGIUM

Supplement 1

Descriptive Micrometeorology

R. E. MUNN

DESCRIPTIVE MICROMETEOROLOGY

R. E. Munn

METEOROLOGICAL SERVICE OF CANADA
TORONTO, ONTARIO
CANADA

1966

ACADEMIC PRESS   New York and London

ACADEMIC PRESS INC.
111 Fifth Avenue, New York, New York 10003

United Kingdom Edition published by
ACADEMIC PRESS INC. (LONDON) LTD.
Berkeley Square House, London W.1

LIBRARY OF CONGRESS CATALOG CARD NUMBER: 65-26406

Fourth Printing, 1971

PRINTED IN THE UNITED STATES OF AMERICA

Foreword

This review is a natural outgrowth of a major theme dominating the symposium published in Volume 6 of *Advances in Geophysics*. There was much favorable response to that volume. It presented many views and new findings in juxtaposition and was a deviation from our practice of presenting a theme in perspective. For part of the field this volume has that perspective as an aim.

Micrometeorology has grown, both in its theoretical and practical aspects, at a rapid rate, but because of its varied applications the literature is scattered widely through the periodicals. Much of the work has been sponsored by various government agencies and has resulted in numerous reports, prepared in limited editions, which are often designed to bury rather than to publish research results. Much excellent work deserves a better fate, but in many cases not even an excerpt finds its way into the regular serial publications. Micrometeorology is probably not the only field that has suffered from this practice, which is detrimental to scholarship, inflicted upon it. Dr. Munn's diligence in reviewing this odd assortment of material is to be greatly commended.

A new look at micrometerology is most timely. Aside from forecasting, this is undoubtedly the field in which the meteorologist can have the greatest practical impact on society. Agriculture, forestry, air pollution, city planning, and pest control are just some of the areas in which micrometeorological knowledge is an indispensible tool. It is also the field in which controlled interference into the natural atmospheric events can be exercised with at least approximately predictable results. In a world which will be more and more limited by its land and water resources and in which health is menaced by contamination of the atmosphere, the knowledge gained should not be disregarded. The review also clearly points out where this knowledge is deficient and thus can offer stimulation to fill the gaps.

H. E. LANDSBERG
J. VAN MIEGHEM

Preface

Micrometeorology is concerned with the surface boundary layer, that thin slice of atmosphere extending from the ground up to a height of approximately 50 meters. In order to study this layer, one should consider interactions with the underlying surface and with the immediately adjoining part of the atmosphere above (the planetary boundary layer).

It is difficult to find a generally acceptable definition of micrometeorology. I would like it to include any atmospheric scale of motion in which the Coriolis force of the earth's rotation can be neglected. A sea breeze, for example, would then be a microphenomenon at first, a mesoscale circulation later in the day, and a macroscale feature if it were to become a continental monsoon.

I have given considerable attention to the properties of the underlying medium and have written separate sections on the micrometeorology of soil, short vegetation, forest, water, ice, snow, and built-up urban surfaces. To be consistent, there also should have been a section on the properties and influence of the planetary boundary layer above; however, a first draft proved to be largely a summary of Professor H. Lettau's work, and undoubtedly he would prefer to write his own survey.

I have been subjective in my choice of topics and references and have tried to avoid producing a catalogue of everything that has already been written. A primary objective has been to provide the reader with a facility for following comfortably and enjoyably all the exciting, individual papers to be published in the next few years. Some unrelated material is therefore included with this criterion in mind.

This survey has been used as the outline for a course in micrometeorology given at the University of Toronto. Although most of the students are candidates for an M.A. degree in meteorology, no particular acquaintance with the atmospheric sciences is assumed or required. Students in related disciplines are encouraged to register. No other work proved to be suitable for this course.

I believe there is a general need at this time for a broad descriptive survey of recent advances in micrometeorology. In the last ten years there has been a remarkable increase in the number of papers published in micrometeorology and in the number of journals and scientific report series in which these papers are to be found. The interest is world-wide and is not limited to those with formal training in meteorology; in fact, the meteorologist is now rather in the minority. Many people—chemists, engineers, geographers, botanists, hydrologists, health physicists, glaciologists, town planners, limnologists, oceanographers, air pollution control officers, foresters, and ecologists—have become interested in the surface boundary layer for one reason or another. Some of my friends have become involved in the science of the atmosphere indirectly or accidentally. I like to call them "back-door micrometeorologists," a term not intended to be derogatory; in fact, the expanding cross-fertilization of scientific disciplines has brought new vigor and some extraordinary new talent to bear on the problems of the surface boundary layer.

I wish to acknowledge my indebtedness to a number of individuals: Professor D. Portman introduced me to micrometeorology at the University of Michigan; Professor G. Gill of the same university made me aware of the importance and intricacies of micrometeorological instruments. From the many who assisted me with discussions or proofreading, I would like to single out for special recognition Mr. E. J. Truhlar of the Meteorological Service of Canada and the students of my University of Toronto class of 1964–1965.

I have been fortunate in having at my disposal the comprehensive collection of journals of the Meteorological Service of Canada; I am indebted in particular to its Chief Librarian, Miss M. Skinner. The average university library is deficient in its selection of journals with micrometeorological import. This survey should therefore prove useful to those without the facilities (or the time) to study the literature.

Toronto, Canada

R. E. MUNN

Contents

FOREWORD   v

PREFACE   vii

1. The Earth-Atmosphere Boundary

1.1. The Scope of Micrometeorology   1
1.2. The Energy Balance at the Earth-Atmosphere Boundary   2
1.3. Models in Micrometeorology   6
1.4. Micrometeorology and Microclimatology   6

2. Short-Wave Radiation at the Earth's Surface

2.1. The Spectrum of Radiation   8
2.2. The Effect of Temperature on Radiant Energy   8
2.3. The Energy from the Sun at the Outer Edge of the Atmosphere   9
2.4. Depletion of Solar Energy by the Atmosphere   9
2.5. Optical Air Mass   12
2.6. An Illustrative Example   13
2.7. Reflection by the Earth's Surface Q_R 14
2.8. The Estimation and Measurement of Q_T and Q_R 15

3. Long-Wave Radiation at the Earth's Surface

3.1. Long-Wave Radiation from the Earth's Surface $Q_{L\uparrow}$ 17
3.2. Long-Wave Radiation from the Sky $Q_{L\downarrow}$ 18
3.3. Radiative Flux Divergence   19
3.4. Measurement of Long-Wave and Net Radiation   21

4. Soil Temperature and Moisture

4.1. Surface Temperature   23
4.2. Subsurface Soil Temperatures   25

4.3. Moisture in Bare Soil   27
4.4. Evapotranspiration   29
4.5. The Lysimeter   32

5. Soil Heat Transfer

5.1. Heat Transfer in a Solid   33
5.2. The Fourier Heat Conduction Equation in One Dimension   34
5.3. Experimental Methods   36
5.4. Some Estimates of Soil Heat Flux   37
5.5. Soil Moisture Flux   37

6. Air Temperature and Humidity near the Earth's Surface

6.1. Factors Influencing Air Temperatures   42
6.2. Diurnal and Annual Patterns of Air Temperature Differences   45
6.3. Precipitation and Fog   48
6.4. Humidity near the Earth's Surface   49
6.5. The Measurement of Temperature and Mixing Ratio   49

7. Wind Flow over Homogeneous Surfaces

7.1. The Essential Problem   53
7.2. Dimensional Analysis and Similarity Theory   54
7.3. Viscosity and Shearing Stress   56
7.4. The Vertical Wind Profile in the Absence of Buoyancy   59
7.5. The Vertical Wind Profile in a Nonadiabatic Atmosphere   61
7.6. The Measurement of Mean Wind and Surface Shearing Stress   65

8. Turbulence over Homogeneous Surfaces

8.1. The Nature of Turbulence   66
8.2. Some Definitions   67
8.3. The Problems of Normality and Intermittency in Shear Zones   69
8.4. The Spectrum of Turbulence   70
8.5. The Kolmogorov Similarity Theory   71
8.6. The Effect of Sampling and Smoothing Times   73
8.7. Correlation Coefficients and the Scale of Turbulence   74
8.8. Cross-Spectrum Analysis   76
8.9. Shearing Stress in Terms of Eddy Fluctuations   77

8.10. The Lagrangian Reference Frame 78
8.11. The Measurement of Turbulence 79

9. Turbulent Transfer of Heat from Homogeneous Surfaces

9.1. The Assumption of Constant Vertical Heat Flux 81
9.2. The Monin-Obukhov Length and the Richardson Number 81
9.3. The Ratio of Diffusivities K_H/K_m 83
9.4. Daytime Turbulent Heat Fluxes 84
9.5. Nighttime Turbulent Heat Fluxes 86
9.6. Viscous Dissipation and the Diabatic Wind Profile 86
9.7. The Eddy Correlation Method for Measuring Heat Flux 88
9.8. The Effect of Radiative Flux Divergence on Heat Transfer 89

10. Evaporation from Homogeneous Surfaces

10.1. The Evaporation Process 92
10.2. Some Formal Relations 93
10.3. The Ratio of Diffusivities 94
10.4. Some Recent Experimental Data 95
10.5. The Eddy Correlation Method for Measuring Evaporation 96
10.6. Some Practical Considerations 96
10.7. Measurement of Temperature and Water Vapor Fluctuations 98

11. Wind Flow around Obstacles

11.1. The Surface of the Earth 99
11.2. Wind Flow around a Cylinder 99
11.3. Wind Flow around Irregular Objects 101
11.4. The Energy Balance of an Enclosed Area 104
11.5. The Effect of a Tower on Wind Measurements 104

12. Transitional Zones and States

12.1. Introduction 107
12.2. The Fetch Required to Achieve Steady State Conditions Downwind
from an Obstacle 108
12.3. The Effect of a Discrete Change in Roughness 108
12.4. Advection 112
12.5. Transitional States 115

13. Atmospheric Pollution

13.1.	The Meteorological Problem	118
13.2.	A Diffusion Model from Probability Theory	119
13.3.	Taylor's Theorem	121
13.4.	The Pasquill Diffusion Model	122
13.5.	The Effect of a Lapse Rate on Plume Behavior	123
13.6.	Effective Stack Height	124
13.7.	Aerodynamic Downwash around a Building	125
13.8.	Transitional Zones and States	127

14. The Air over Bare Ground

14.1.	Models and Reality	129
14.2.	The Energy Balance of a Dry Surface	129
14.3.	The Energy Balance of Moist Ground	131
14.4.	The Effect of Fences and Hedges	132

15. The Air over Snow and Ice Surfaces

15.1.	Introduction	133
15.2.	The Radiation Balance of Snow and Ice Surfaces	133
15.3.	Heat Flux and Heat Storage within Snow and Ice	135
15.4.	Wind Profiles over Snow and Ice	138
15.5.	Temperature Profiles and Vertical Heat Transfer	138
15.6.	Humidity Profiles and Latent Heat Transfer	141

16. The Energy Balance of a Plant Cover

16.1.	The Air over a Short Grass Surface	143
16.2.	The Energy Balance of a Leaf	144
16.3.	The Radiation Balance of a Plant Cover	145
16.4.	Profiles within a Plant Cover	147
16.5.	Profiles above Plant Covers	148
16.6.	The Energy Balance within a Plant Cover	149
16.7.	An Alternative Notation for Fluxes	151
16.8.	Carbon Dioxide Profiles and Fluxes	153

17. Forest Meteorology

17.1.	The Forest: An Active Meteorological Region	155
17.2.	The Radiation Balance of a Forest	155

17.3. Soil Temperature and Moisture 157
17.4. Forest Temperatures 158
17.5. Winds in the Forest 161
17.6. Humidity in the Forest 161
17.7. Heat Storage within Trees 162
17.8. The Energy Balance of a Forest 163
17.9. Additional Remarks 166

18. The Air over Oceans and Large Lakes

18.1. Introduction 167
18.2. Some Physical Properties of Oceans and Lakes 167
18.3. The Radiation Balance of Oceans and Lakes 169
18.4. Temperature and Humidity near the Water Surface 170
18.5. Wind Flow over Water 174
18.6. Heat Storage and Horizontal Advection in Water 176
18.7. Energy Balance Calculations of a Lake or Ocean 177
18.8. Turbulence over Water 178

19. Land and Sea Breezes

19.1. Introduction 179
19.2. Land and Sea Breezes during Light Geostrophic Winds 180
19.3. Sea Breezes When a Geostrophic Wind Is Blowing 183
19.4. A Mathematical Model of the Sea Breeze 184
19.5. Humidity Profiles Associated with Sea Breezes 187
19.6. The Micrometeorology of Small Islands and Lakes 187

20. The Air in Valleys

20.1. The Importance of Valley Meteorology 189
20.2. Valley Influences during Strong Geostrophic Winds 189
20.3. The Radiation Balance in Hilly Country 190
20.4. Local Wind Flows in Valleys during Light Geostrophic Winds 191
20.5. Temperatures and Pollution in Valleys 194

21. The Air over a City

21.1. Introduction 195
21.2. The Radiation Balance of a City 196

21.3. Conductive Heat Flux Q_G 199
21.4. The Heat Generated by a City   199
21.5. City Temperatures   199
21.6. The Humidity in a City   204
21.7. Winds in a City   204
21.8. The Effect of Parks and Greenbelts   205
21.9. Conclusion   207

22. The Modification of Local Weather

22.1. Introduction   208
22.2. Changes in the Radiation Balance   208
22.3. Changes in Soil Heat Flux Q_G 210
22.4. Changes in Humidity   211
22.5. Changes in Wind Patterns   212
22.6. Other Examples of Micrometeorological Weather Modification   215

List of Symbols   217

References   221

AUTHOR INDEX   237
SUBJECT INDEX   242

1. The Earth-Atmosphere Boundary

1.1. The Scope of Micrometeorology

The atmosphere is an important natural resource. It protects the earth from cosmic radiation and meteorites; it transfers heat from equator to pole and water from ocean to continent; and it provides the air we breathe.

The atmosphere is rarely at rest. The motion is mainly in a horizontal direction but the individual fluid elements show a strong tendency to move in spirals. A whirl or eddy may be as large as a continent (*macroscale*), the size of a few thunderstorms (*mesoscale*), or smaller than a city (*microscale*). Large eddies contain a multitude of smaller eddies ranging in size from kilometers to millimeters. The outline of an individual whirl is poorly defined and is continually changing.

The primary source of energy for atmospheric motion is the sun. World-wide weather patterns (macroscale) result from unequal heating of the earth by the sun. Contributing influences are the deflecting force of the earth's rotation (called the *Coriolis force*), the uneven distribution of oceans and continents, and the location of mountain ranges. Large-scale weather features are revealed by a network of observing stations spaced at intervals of 20 km or so.

Mesometeorological patterns are studied by means of radar or with observing stations spaced at intervals of about 1 km. In micrometeorology, the scale of interest is limited to a few square kilometers, and the Coriolis force can usually be neglected.

Micrometeorology is restricted in depth to the lowest hundred meters of the atmosphere, the *surface boundary layer*. Although this comprises a relatively small fraction of the atmosphere, it is important for two reasons. In the first place, plants, animals, and man live in the boundary layer. A knowledge of micrometeorology is therefore useful in agriculture, hydrology, forestry, and public health, to cite

1

only a few examples. Second, many of the great transformations in the atmosphere take place at its lower boundary. On cloudless days, for example, solar radiation passes through the atmosphere with little reduction in strength; it is only upon reaching the surface of the earth that the sun's rays are converted into heat. Furthermore, evaporation, evapotranspiration, convection, and frictional wind drag largely originate at the earth-atmosphere interface.

There are, of course, important links connecting macro-, meso-, and micrometeorology. Energy is continually being transferred up and down the scales of atmospheric motion. For example, local evaporation contributes in a few hours or days to the generation of storms hundreds or thousands of kilometers away. However, the links are not yet well defined and in any event are beyond the scope of this review.

What scale separates micro- from mesometeorology? It would be useful to have mutually exclusive definitions of both terms but there is at present a rather wide zone of overlap. Valley winds, for example, are considered as both micro- and meso-phenomena; as might be expected, the point of view depends in part upon the size of the valley.

In the last decade there has been increasing emphasis on conservation and management of natural resources. In the case of the atmosphere, modification of local weather is a reality. It is practiced, often unwittingly, in three general ways:

1. By changing the radiation properties of a surface, e.g., by ploughing a field or by replacing forest with city.
2. By introducing or removing local obstacles to the air flow.
3. By changing the water balance of the surface, e.g., drainage of swamps or irrigation of crops.

These factors will be considered in subsequent chapters.

1.2. The Energy Balance at the Earth-Atmosphere Boundary

The principle of conservation of energy states that all gains and losses of energy at the surface of the earth must balance. The primary components are short-wave radiation from the sun, long-wave radiation from the earth and sky, transfer of heat through the ground, transfer of heat through the air, and the contribution of latent heat of evaporation or condensation. Other sources of energy can usually be neglected. The conservation principle can be expressed as a very general equation applicable at any instant in time:

$$(1.1) \qquad Q_T - Q_R + Q_{L\downarrow} - Q_{L\uparrow} = \pm Q_N = \pm Q_G \pm Q_H \pm Q_E$$

where Q_T is the short-wave radiation from sun and sky, assumed by convention to be positive. It is absent at night.

Q_R is the short-wave radiation reflected from the earth. It is absent at night.

$Q_{L\downarrow}$ is the long-wave radiation received by the surface from the atmosphere.

$Q_{L\uparrow}$ is the long-wave radiation emitted by the surface.

Q_N is the net all-wave radiation. A gain of energy by the surface is positive.

Q_G is the transfer of heat through the ground. A downward flow of heat is positive.

Q_H is the turbulent transfer of sensible (i.e., not involving changes of state) heat to the atmosphere. An upward flow is positive.

Q_E is the contribution of latent heat of evaporation and evapotranspiration. An upward flow of water vapor (evaporation) is positive; a downward flow (condensation) is negative.

The expression "energy balance" does not imply that individual components in equation (1.1) are necessarily constant with time. This latter special case is called a *steady state condition*. The conventional unit of rate of energy transfer or *flux* is the langley per unit time, the *langley* (ly) being defined as one gram calorie per square centimeter. The term Q_E is related to water loss by equation (1.2):

$$(1.2) \qquad\qquad\qquad Q_E = LE$$

where L is the latent heat of evaporation (published in standard tables) and E is the rate of evaporation (mass per unit area per unit time). Thus it is an easy conversion from water loss to the corresponding exchange of latent heat.

During the daytime there is a gain of radiant energy by the surface (positive Q_N). Part of the surplus moves downward into the soil, part is transferred upward to the air, while the remainder is used in evaporational cooling. The terms Q_G and Q_H are real flows of heat. Evaporational cooling, on the other hand, takes place only at the surface. Although there is a resulting upward flow of water vapor, the heat exchange is felt through a compensating reduction in the magnitude of Q_G and Q_H. It is possible on some occasions (such as irrigation of a desert) for evaporational cooling to be greater than the available radiant energy Q_N; the fluxes of Q_H and/or Q_G must then reverse directions.

At night there is a net loss of radiation by the surface (negative Q_N). This is balanced by heat flows upward through the ground, downward from the air, and at times by condensational heating (dew formation). Typical day and night conditions are shown schematically in Fig. 1.

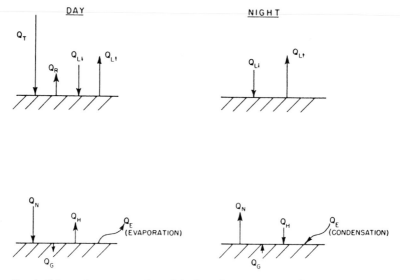

Fig. 1. Schematic representation of the heat fluxes at the earth's surface by day and by night.

Because of instrumental difficulties, it is not feasible to take measurements exactly at the surface, if in fact a surface can be defined over rough terrain. Observations are therefore taken a short distance from the interface. A graph showing the variation with height of wind, temperature, or humidity is called a *vertical profile*. At this point, a fundamental hypothesis of heat transfer theory should be mentioned. If the temperature is constant everywhere in an insulated medium, there will be no heat transfer. It is only when temperature differences occur that heat flows arise. Vertical temperature profiles will therefore provide some clues about the behavior of Q_G and Q_H. Similarly, evaporation will take place only when water vapor differences exist. Hence, vertical profiles of soil moisture and of atmospheric humidity should yield information about the magnitude of Q_E. These ideas will be developed in Chapters 5, 9, and 10.

Equation (1.1) is often used to estimate the magnitude of one type of energy transfer from a knowledge of the others. In general, the radiation terms can be measured directly and with greater accuracy than any component on the right side of equation (1.1). Over a land surface, Q_G measurements are next in accuracy; over an ocean or lake, Q_G is often the unknown term. Finally, Q_H and Q_E present many experimental and theoretical difficulties.

Early attempts at estimating Q_H and Q_E were based on a molecular analogy. It was, of course, recognized that the contribution of molecular conduction to the transfer of heat and water vapor was trivial in the atmosphere. The exchange processes are dominated by a wide range of turbulent eddies. However, Bowen [1] introduced the *Bowen ratio R*:

(1.3) $$R = Q_H/Q_E$$

Then the energy balance equation becomes

(1.4) $$Q_E = \frac{Q_N - Q_G}{1 + R}$$

or

(1.5) $$Q_H = \frac{Q_N - Q_G}{1 + 1/R}$$

The assumption was made that the relative efficiencies with which heat and water vapor are transferred are the same for turbulent as for molecular processes. The validity of the assumption will be considered later. However, if the ratio R can be estimated with any accuracy, then Q_E and Q_H may be obtained from experimental measurements of net radiation and of soil heat flux. At this stage it can be seen that if Bowen's ratio is much smaller than unity, the relative error in an estimate of Q_E from equation (1.4) will be much smaller than the error in the ratio itself. Similarly, if R is large, as in a desert, equation (1.5) will provide good estimates of Q_H.

1.3. Models in Micrometeorology

It is common in the geophysical sciences to describe nature with idealized physical and mathematical models. The important factors in a particular situation are isolated for study; other influences that make it impossible to describe the relationships mathematically or which lead to equations that cannot be solved are excluded from the analysis. For example, the compressibility of air is often neglected. The results are then compared with behavior in the real atmosphere.

Models serve three main purposes:

1. They develop a qualitative understanding and appreciation for the physical mechanisms.
2. They can be tested experimentally. Consistency of results must lead to greater confidence in the model and the measurement techniques.
3. They can be used for predicting the behavior of the variables.

It must be emphasized, however, that a physical model is not a law of nature. The basic assumptions must be clearly recognized before intelligent use may be made of the resulting predictions. There is unfortunately a tendency for physical formulas appearing in engineering handbooks to be applied quite out of context.

Probably the most important assumption in micrometeorology is that at a height of about 500 to 1000 meters there is an undisturbed constant air flow called the *geostrophic wind*. The results are therefore applicable only for periods when the meso- and macroscale features remain unchanged and there has been time for equilibrium to be reached in the entire 500-meter layer. Micrometeorological conditions associated with a changing geostrophic wind have not yet been investigated.

1.4. Micrometeorology and Microclimatology

There is unnecessary confusion about the difference between micrometeorology and microclimatology, a situation that is not helped by the definitions given in dictionaries and glossaries.

The micrometeorologist is concerned with instantaneous fluxes of heat, matter, and momentum under idealized conditions. He prefers to work over an "infinite plane" but he is interested in discontinuities such as a lake shore or the edge of a forest. In any event he is willing to compute evaporation rates or heat fluxes at some fixed point for some fixed instant. In practice, the fixed instant is replaced by a period lasting from 10 min to an hour or so; intervals of time are usually selected for which the data exhibit no trends (steady state conditions), although there may be random fluctuations.

The agriculturist and the hydrologist are not interested in such detail but instead require daily, monthly, or even seasonal estimates of evaporation or heat flows over large nonuniform surfaces perhaps the size of an entire drainage basin. These are the problems that confront the microclimatologist.

Despite the difference in emphasis, microclimatology must ultimately be an extrapolation in time and space of micrometeorology. At the present time this is not always possible because of the cost of instruments and because of the lack of fundamental knowledge of exchange processes over irregular surfaces and under conditions of variable geostrophic wind. A number of simplified equations have therefore been used, some of which have proved to be quite satisfactory for engineering purposes.

2. Short-Wave Radiation at the Earth's Surface

2.1. The Spectrum of Radiation

Any body not at absolute zero transmits energy to its surroundings by radiation. The energy is in the form of electromagnetic waves traveling with the speed of light and requiring no intervening medium.

Wavelengths are measured in terms of the micron (1 μ = 10^{-4} cm) or the Ångström (1 Å = 10^{-8} cm). There is a wide range or spectrum of wavelengths extending from the very short cosmic and x-rays ($< 0.29 \mu$), through the ultraviolet ($0.29 - 0.40 \mu$) and the visible ($0.40 - 0.63 \mu$) to infrared, microwave, and radio waves ($> 0.63 \mu$). A diagram displaying the energy emitted by a body at different wavelengths (or wavelength bands) is known as a *spectrum*.

2.2. The Effect of Temperature on Radiant Energy

A *blackbody* is one that absorbs all of the electromagnetic radiation striking it. Usage varies and sometimes the term is restricted to specific wavelength bands. In any event the definition does not imply that the object must be black in color. For example, snow is an excellent blackbody in the infrared part of the spectrum.

For a perfect all-wave blackbody, the intensity of radiation and the wavelength distribution depend only on the absolute temperature of that body. The Stefan-Boltzmann law applies:

(2.1)
$$F = \sigma T^4$$

where F is the flux of radiation (ly/min),
T is absolute temperature,
σ is equal to 0.813×10^{-10} cal/cm² min(°K)⁴.

It can also be shown that the wavelength of maximum energy λ_m is inversely proportional to absolute temperature.

(2.2) $$\lambda_m = \alpha/T$$

where $\alpha = 0.288$ cm °K.

Although the sun and earth do not behave as perfect blackbodies, experimental evidence indicates that equations (2.1) and (2.2) may be applied as good approximations. The wavelengths of maximum energy for Q_T and $Q_{L\uparrow}$ are about 0.47 and 10 μ, respectively, in general agreement with the prediction of equation (2.2). Because of the great difference in temperature between sun and earth, the spectra of solar and terrestrial radiation do not overlap to any appreciable extent, providing a convenient separation of the two streams of radiant energy. Whereas 99 % of solar energy lies between wavelengths of 0.15 and 4 μ (largely in the visible), 99 % of terrestrial radiation lies between 4 and 120 μ (in the infrared).

2.3. The Energy from the Sun at the Outer Edge of the Atmosphere

When the earth is at its mean distance from the sun, the flux of solar radiation falling on a surface normal to the solar beam at the outer edge of the atmosphere is 1.99 ± 0.02 ly/min [2]. This is known as the *solar constant* I_0. Because of latitude, season, and time of day, there is a reduction in energy available to a horizontal surface, i.e., a surface slanted to the rays of the sun. The resulting value (called I_h) may be obtained by solving an equation in spherical trigonometry for which Mateer and Godson [3] have published a nomogram.

Figure 2 [4] presents the global daily distribution, assuming an earlier estimate of 1.94 ly/min for the solar constant. The features of interest are the sharp summer peak at the poles, the even distribution at the solstice over the summer hemisphere, and the rapid changes in spring and autumn in arctic and antarctic latitudes. The radiant flux given in Fig. 2 is the primary source of energy for atmospheric motions.

2.4. Depletion of Solar Energy by the Atmosphere

The atmosphere weakens the solar beam by absorption, scattering, and reflection. If light only reached the ground in a straight line from the sun, the sky would be a black curtain covered with stars.

Energy in wavelengths less than 0.29 μ does not reach the ground, largely because of absorption by ozone in the upper atmosphere. This amounts to a loss of slightly more than 5 % of the incoming solar beam, but it fortunately protects man from cosmic and the more intense ultraviolet radiation. Water vapor and carbon dioxide absorb

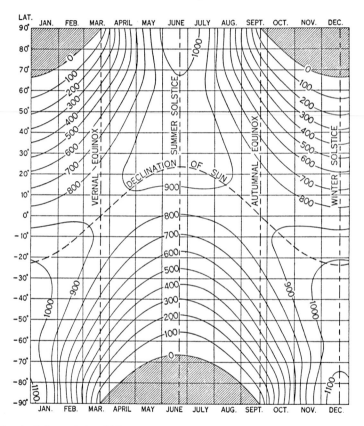

FIG. 2. Solar radiation falling on a horizontal surface at the outside of the atmosphere in ly/day. Shaded areas represent regions of continuous darkness [4].

solar energy at a number of discrete wavelengths. Clouds have a variable but sometimes very large effect on the depletion of energy from all parts of the solar spectrum.

Air molecules, water vapor, and very small particles scatter the sun's rays. Scattering varies inversely as the fourth power of the wavelength, the Rayleigh law for very small particles. The sky appears blue because there is more scattering of blue light than there is of other colors in the longer wavelengths.

Large solid particles in the atmosphere reflect rather than scatter light, all wavelengths being affected equally. The blue color of the sky therefore turns to white with increasing pollution.

Because a sensor at ground level cannot distinguish between scattered and reflected light, the two are often considered together and are called *diffuse* solar radiation (the total short-wave energy received in the shade). When skies are clear, the ratio of direct to diffuse solar

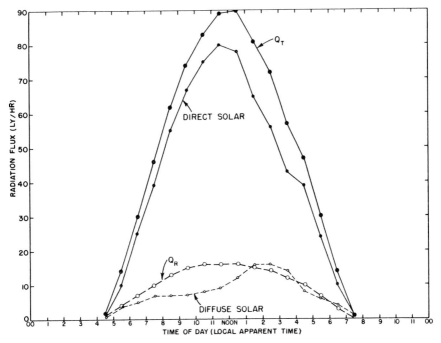

FIG. 3. Solar radiation components at Toronto, Canada on June 27, 1962 over grass.

radiation is about 6 to 1 when the sun is directly overhead, falling to about 2 to 1 when the sun is 20° above the horizon.

Figure 3 illustrates the relative magnitudes of the solar components on a sunny day at Toronto, Canada; the reflected radiation Q_R is also included. An interesting feature is the increase in diffuse solar energy in the afternoon associated with the development of some high cirrus clouds.

2.5. Optical Air Mass

The energy received at the earth's surface Q_T depends greatly upon the path-length of the beam through the atmosphere. This can vary from the normal depth of one atmosphere with the sun overhead to a value 40 times greater when the sun is near the horizon, i.e., there will be 40 times as many molecules and solid particles to deplete the sun's energy.

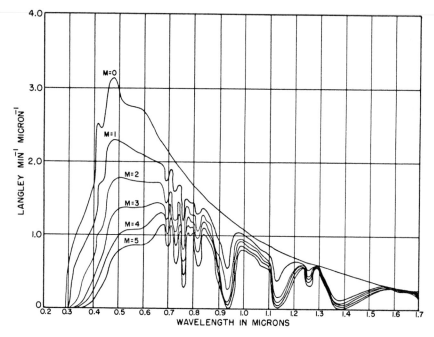

FIG. 4. Spectral distribution of solar energy falling on a normal surface for six optical air masses [5].

The *optical air mass m* is a measure of the length of path through the atmosphere traversed by rays from a celestial body, expressed as a multiple of the path-length to sea level for a source at the zenith (directly overhead). The optical air mass is zero at the outer edge of the atmosphere. Figure 4 [5] displays the spectral distribution of solar energy falling on a normal surface for six optical air mass values, assuming a cloudless atmosphere containing 20 mm of precipitable water vapor, 2.8 mm of ozone, and 300 particles/cm^3 of dust. The sea level pressure is assumed to be 1013.25 mb. The total area under the upper curve ($m = 0$) is equal to the solar constant. The selective nature of the energy reduction (called *attenuation*) by a cloudless atmosphere is illustrated for optical air masses 1 to 5.

For monochromatic light (single wavelength), the attenuation through a medium is given by Beer's law:

$$(2.3) \qquad\qquad I = I_0 \exp(-am)$$

where I and I_0 are final and initial energy intensities,

a is called the *extinction coefficient*,

m is optical air mass,

I/I_0 is defined as *transmissivity*,

am is defined as *optical density*.

Although Beer's law is valid only for a single wavelength, it is some-times used as an empirical approximation for solar energy attenuation. It has also been applied to the reduction of net radiation within a crop (Chapter 16) or in connection with light penetration in snow, ice, or water (Chapters 15 and 18). The extinction coefficient varies with optical air mass, wavelength, and the nature of the attenuation (by scattering, absorption, or reflection). Despite these complications, it is sometimes useful to know that a stream of energy is diminishing more or less exponentially. Measurements at two levels will then permit approximate interpolation or extrapolation.

2.6. An Illustrative Example

Table I [6] gives comparative values of solar radiation falling on a horizontal surface at Toronto, Canada under various atmospheric conditions. The numbers in brackets express the various totals as percentages of I_h. The observed values of Q_T, of course, include diffuse as well as direct solar energy. Table I is included merely for illustration, and it is not implied that results would be similar in other parts of the world.

The computed values for a clear atmosphere with water vapor are based on a theoretical method [7] assuming average seasonal amounts of water vapor. Atmospheric scattering and absorption by air free of water vapor is about 7% except in December (15%) when the

TABLE I. Solar radiation falling on a horizontal surface at Toronto, Canada (ly/day).

	Sept. 15	Dec. 15	Mar. 15	June 15
No atmosphere, I_h	675(100)	264(100)	595(100)	971(100)
Clear atmosphere with no water vapor	625(93)	224(85)	548(92)	911(94)
Clear atmosphere with water vapor, Q_T				
(a) Computed	525(78)	207(79)	490(83)	766(79)
(b) Observed	480(71)	180(68)	460(77)	740(76)
Average observed Q_T for all days (both clear and cloudy)	350(52)	85(32)	280(38)	530(55)

optical path is considerably longer. Water vapor reduces the total solar radiation by 15% in summer and autumn but only 6 to 9% in winter and spring because the average water vapor content is smaller in the latter seasons. Observed clear sky values of Q_T are lower than computed values, mainly because of the polluted air in the Toronto area. In winter and spring the cloudy nature of the Great Lakes climate is shown by the low values of Q_T when all days are averaged.

2.7. Reflection by the Earth's Surface Q_R

The *albedo A* of a surface is the ratio of the flux of solar radiation reflected and scattered by a surface to the flux incident upon it. It follows that

$$(2.4) \qquad A = Q_R/Q_T$$

The solar radiation available for energy transformations at the earth's surface is therefore $(1 - A)Q_T$.

The definition of albedo varies throughout the meteorological literature. In some applications albedo is restricted to the visible wavelengths while on other occasions the definition is widened to include terrestrial long-wave radiation. Albedo is sometimes called *reflectivity* but the latter term more properly refers to the reflected–incident ratio for a specific wavelength and implies mirrorlike reflection. Reflection does vary with wavelength, e.g., grass is green because it reflects much of the green light and absorbs most of the energy in the other colors. Both albedo and reflectivity vary with solar elevation, particularly over a water surface, being greatest when the sun is near the horizon. Hence, the albedo for diffuse short-wave sky radiation is slightly greater than for direct solar. Some typical values of albedo [as defined by equation (2.4)] are given in Table II.

TABLE II. Typical values of albedo [8].

Surface	Albedo, %
Snow, fresh	70–95
Snow, several days old	70
Dry dune sand	37
Moist dune sand	24
Wet grass in sun	33–37
Wet grass, no sun	14–26
Dry grass	15–25
Forest: pine, fir, and oak	10–18
Ocean	2–7

In general, the low albedo of soils and water results in a high absorption of solar energy at the ground; the high albedo of snow leads to a low absorption. A cloud cover tends to mask the direct rays of the sun so that albedo changes little with the sun's altitude. Typical values measured at Toronto, Canada over a grass surface on a clear and on a cloudy day are given in Table III [6].

TABLE III. Albedo of a grass surface (in per cent) at Toronto, Canada.

Hour (true solar time)	07	08	09	10	11	12	13	14	15	16	17
Sept. 9, 1961 (clear)	25.6	22.3	22.5	21.3	20.8	20.6	20.8	21.4	22.8	24.2	25.3
Sept. 26, 1961 (cloudy)	19.7	19.1	20.2	17.5	19.4	17.6	18.1	19.3	20.0	19.7	28.6

The height of the grass (5 cm) tends to cause multiple reflections which prevent the albedo from assuming much larger values toward sunrise and sunset; on the clear day the albedo was only about 5% higher at 07 and 17 hr than it was at noon. On the second day, the high value of 28.6% at 17 hr occurred with partly cloudy skies.

2.8. The Estimation and Measurement of Q_T and Q_R

If the vertical distribution of water vapor in the atmosphere is known by a concurrent radiosonde flight, Q_T may be computed by a simple graphical technique [7] for clear skies and no pollution. In the absence of water vapor information or when clouds are present, many empirical equations have been proposed, e.g., by Budyko [9]. However, it should be emphasized that such equations can only be used to estimate long-term averages of Q_T. Daily or hourly amounts of radiation will not in general obey the same empirical equations, e.g., reflections from clouds and snow surfaces may cause large short-term fluctuations in Q_T. In the absence of measurements, the albedo values of Table II may be used to provide estimates of the energy component Q_R.

Solar radiation falling on a horizontal surface can be measured directly with an accuracy of about 5% [10]. Extensive use is made of the Eppley 180° pyrheliometer in North America and the Kipp Solarimeter in Europe. Details of construction, spectral response, and linearity have been reviewed by Latimer [11]. Diffuse short-wave sky radiation may be obtained by shading the sensor from direct sunlight. In addition, the instruments may be used in the inverted position to measure the reflected component Q_R. In most applications the instruments are mounted at a height of about 1.5 meters above ground but when uneven terrain is being studied, it is sometimes preferable to seek integrated areal estimates of Q_R from aircraft, e.g. [12].

3. Long-Wave Radiation at the Earth's Surface

3.1. Long-Wave Radiation from the Earth's Surface $Q_{L\uparrow}$

In the long-wave part of the spectrum, the earth behaves very much like a perfect blackbody. The slight experimental correction is given by *emissivity* ε defined as the ratio of emission of the surface to the emission of a perfect blackbody at the same temperature and wavelength. For most surfaces, ε extends from slightly below 1.00 to 0.85. Pure water has an average emissivity of 0.985 while fluffy snow is one of the best natural blackbodies in the infrared with $\varepsilon > 0.99$.

Most surfaces behave approximately as *graybodies*, i.e., emissivity is independent of wavelength. With that assumption, the upward long-wave radiation is given by equation (3.1).

$$(3.1) \qquad Q_{L\uparrow} = \varepsilon\sigma T^4 + (1 - \varepsilon)Q_{L\downarrow}$$

The net upward long-wave radiation is equal to the radiation from the surface plus the small fraction of downward sky radiation that is not absorbed by the surface.

Equation (3.1) would be useful in determining the surface temperature T if it were not for the fact that it is very difficult to estimate emissivity experimentally. In some applications, therefore, the true surface radiative temperature T is replaced by an *apparent surface radiative temperature* T^*, assuming $\varepsilon = 1$.

$$(3.2) \qquad Q_{L\uparrow} = \sigma T^{*4}$$

The apparent surface temperature usually lies between the true surface temperature and that measured in a standard Stevenson screen.

3.2. Long-Wave Radiation from the Sky $Q_{L\downarrow}$

In the absence of an atmosphere, upward long-wave radiation would be lost directly to space. Fortunately, the atmosphere absorbs about 70% of $Q_{L\uparrow}$ (during clear skies), reradiating it in all directions including a downward stream back to the ground. "Our atmosphere easily admits solar radiation but lets earth radiation out only reluctantly" [13]. This is called the *greenhouse effect*, but the analogy is not correct because a major factor in greenhouse climate is the protection the glass gives against turbulent heat losses.

An important property of sky radiation is that its absorption by gases is not continuous over the spectrum but occurs in a series of discrete lines. In some parts of the spectrum, infrared radiation can move upward relatively freely and be lost to space when skies are clear; the most notable example is the so-called window from 8.5 to 9.5 μ. Radiation in some other wave bands, on the other hand, may be almost completely intercepted by the atmosphere. An example of observed sky radiation spectra for four optical air masses is given in Fig. 5 [14]. Skies were clear and screen temperature varied between

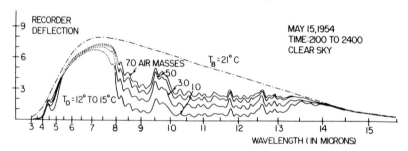

FIG. 5. Spectral distribution of long-wave sky radiation for four optical air masses at Columbus, Ohio [14].

12 and 15°C. The upper envelope is the theoretical blackbody curve for a temperature of 21°C.

Many attempts have been made to obtain estimates of $Q_{L\downarrow}$ from radiosonde measurements of water vapor and temperature. A number of radiation charts has been constructed, the recommended one being that of Yamamoto [15]. The charts differ in the approximations used to replace discrete molecular absorption lines by smoothed wave bands, and in the nature of the carbon dioxide corrections. The effects of variable cloudiness and of urban pollution are not included.

Radiation charts are not particularly useful for the micrometeor-ologist because concurrent radiosondė observations are rarely available. It is preferable to measure $Q_{L\downarrow}$ directly with a radiometer.

If skies become overcast, it is usually assumed that the cloud base radiates as a blackbody; $Q_{L\downarrow}$ may then be computed from equation (3.1) assuming a cloud emissivity of unity with sufficient accuracy. If cloud base temperature is higher than that at the ground, the sur-face of the earth will gain energy until equilibrium is reached.

When cloudiness is variable, some allowance must be made for the amount and height of cloud. A number of empirical formulas have been proposed but they cannot be expected to apply except for long-term climatological estimates.

3.3. **Radiative Flux Divergence**

The energy balance, equation (1.1), is to be determined at the inter-face $z = 0$. However, all terms are evaluated on the basis of measure-ments taken a short distance from the surface. It is therefore pertinent to inquire whether the fluxes are constant with height.

Flux divergence occurs when the vertical flux varies with height, regardless of sign. If the upward flux is greater at 2 than at 4 meters, energy is being added to the intervening layer. If all other forms of energy transfer are constant with height, temperatures will rise. Con-versely, if upward flux is less at 2 than at 4 meters, energy is being lost by the layer and the air will cool. When all kinds of heat exchange are constant with height, temperatures are also constant.

The fact that air temperatures do not change with time has some-times been considered as an indication that flux divergence is absent during the period being studied. This is an unjustified assumption; flux divergences of Q_N and Q_H may be present but in equilibrium with each other. The interpretation of temperature profiles must therefore be approached with great care.

In the absence of fog, short-wave radiative flux divergence is negligible in the lowest few meters of the atmosphere [16]. The height at which Q_T and Q_R are measured is therefore not critical. Such is not the case for long-wave radiation.

Robinson [17] estimated from his radiation chart that the maxi-mum divergence in a 1-meter air layer over short grass was about -0.007 ly/min during the day and 0.002 ly/min at night. For a number of years there were no experimental measurements to check these predictions and long-wave radiative flux divergence was largely ignored. Recent studies by Funk have reopened consideration of the question by micrometeorologists.

Funk [18] used two carefully balanced net radiometers at night (when $Q_N = Q_{L\downarrow} - Q_{L\uparrow}$) to determine flux divergence at heights varying from 0.35 to 7.5 meters over a grass surface. Over a period of 39 nights in Australia, the observed flux divergences were consistently higher than those computed from radiation charts, the ratio varying from 1.3 to 2.8. Funk attributed this to the presence of haze. A comparison was also made of the observed cooling rate and that to be expected solely from flux divergence. The latter was usually greater than the former, the ratio of the two averaging 2.8 over 9 cases. Radiative cooling rates were as high as 12°C/hr.

Another interesting feature was that on several occasions the shapes of the radiative and the actual cooling rate curves were similar, in-phase, wavy traces with periods of about 40 min. An example is given in Fig. 6.

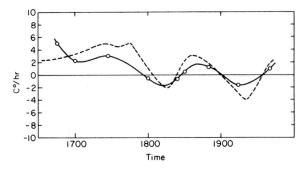

FIG. 6. Comparison of cooling rates at Edithvale, Australia, July 10, 1959. Solid line is the radiative rate averaged over the 1–2-meter layer; dashed line is the actual rate at a height of 1.5 meters [18].

The main conclusions to be drawn from Funk's study are that:

1. Radiative flux divergence is a major cause of air cooling at night.
2. Since actual cooling rates are usually not the same as radiative cooling rates, the eddy transfer component must be variable in the vertical or in the horizontal.

In a later paper Funk [19] presented a numerical method for the computation of radiative flux divergence near the ground from profiles of temperature and water vapor pressure. For one specific case, the numerical value was slightly lower than' the measurement obtained from two balanced net radiometers. Recently, however, Funk [20] and Hamilton [21] have found quantitative disagreement between observed and computed flux divergences, assuming H_2O and CO_2 to be the only absorbers. Funk noted that the actual maximum divergence was often at a height of 3–4 meters and occurred when winds were less than 1 meter/sec. He suggested that a nonuniform vertical distribution of haze might be important and he speculated that there was a maximum concentration of haze at the 3–4 meter level in his experiments. The discrepancies could then be explained.

Because net long-wave radiation is usually only a small fraction of net all-wave radiation in the daytime, few experimental data have been obtained on flux divergence when the sun is shining. Nevertheless, the available evidence suggests that the effect is significant, particularly when the water vapor content is high.[1]

3.4. Measurement of Long-Wave and Net Radiation

Net radiation Q_N can be measured with an accuracy of about 5% with the ventilated Suomi [22] or the polythene-shielded Funk radiometer (the Suomi radiometer cannot be used in rain or prolonged fog). A comparison of the radiometers has been given by Latimer [23].

It is difficult but not impossible to obtain accurate values of $Q_{L\uparrow}$ and $Q_{L\downarrow}$ individually. The upper or lower surface of a radiometer may be shielded with an enclosure blackened on the inside and at known temperature. However, careful attention must be given to calibration and maintenance.

For application of equation (1.1) in energy balance studies, it is neither necessary nor desirable to measure the various radiation components separately; the quantity Q_N can be obtained directly with a net radiometer. In the daytime, long-wave flux divergence will not make a significant contribution to the magnitude of Q_N because of the large short-wave component, although it is important for other reasons (see Chapter 9). The height of the radiometer is therefore not critical. At night, on the other hand, net radiation measurements obtained at a height of 1.5 meters may not be representative of the

[1] Godson (First Canadian Conference on Micrometeorology, Toronto, 1965) has recently clarified many of the problems associated with radiation-chart flux divergence calculations, including the effect of aerosols, and the errors involved in aircraft radiation thermometry for estimating surface temperature.

radiation balance at the interface. Fortunately, the total energy is small during hours of darkness and does not make a large contribution to daily or monthly averages.

In plant growth and health studies, the energy in specific wavelength bands is often required. Within a forest, for example, the wavelengths of use in photosynthesis may be almost completely depleted by the upper canopy; the solar radiation reaching the forest floor is then sufficient in quantity but insufficient in quality. Measurements of this kind can be obtained by using suitable optical filters.

When investigating radiative flux divergence experimentally, it should be remembered that the surface area being sampled increases as the height of the radiometer is increased. When the sensor is z meters above the ground, it receives 90 and 99 % of its upward flux from a surface area with radius $3z$ and $10z$ meters, respectively.

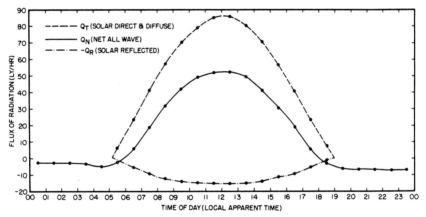

Fig. 7. The radiation components at Toronto, Canada on May 11, 1962 during clear skies over grass.

Some typical clear-day radiation measurements at Toronto, Canada are given in Fig. 7. The net radiant energy to be shared among Q_G, Q_H, and Q_E is at a maximum near noon; at night Q_N values are negative and quite small in absolute magnitude.

4. Soil Temperature and Moisture

4.1. Surface Temperature

Because of the uneven nature of the ground, it is difficult to define precisely the average height of the interface, the reference plane $z = 0$. It is even more difficult to measure its temperature accurately. The largest temperature gradients occur within ± 1 cm of the boundary and may exceed 5°C/cm at times. Even over an ocean there is often a temperature change of 0.5°C in the top millimeter of water because of evaporational cooling.

A glass thermometer or thermocouple is sometimes placed on the ground but there are many difficulties [24]. Air temperatures are measured in the shade and the sensor is usually provided with some form of forced ventilation. This is quite impracticable at $z = 0$. On the other hand, if the soil surface temperature is to be measured, the sensor must be located exactly at $z = 0$ and must have the same thermal properties as its environment. If it protrudes into the air, it will be affected by wind; if it is partially buried in the ground, it will no longer measure surface temperature. Even when an intermediate position is chosen with care, dirt will soon drift over the instrument or the thermocouple will warp and lose contact with the ground. In addition, a glass thermometer will act like a greenhouse and store solar radiation. In general it may be stated that all such instruments, if carefully attended, will yield an accuracy of about 0.5°C at night but will be subject to much larger errors when the sun is shining.

Another method, in theory at least, of determining surface temperature is through equation (3.1). However, emissivity cannot be determined with sufficient accuracy; equation (3.2) is therefore used to obtain the apparent surface radiative temperature T^*. The method is showing considerable promise for charting variations in surface temperature over wide areas by infrared photography from low-flying aircraft. Nevertheless, problems of calibration and of radiative flux divergence remain to be fully resolved.

23

On a much smaller scale, equation (3.2) has been used by Monteith and Sceicz [25]. Some results from three summer days at Rothamsted, England are given in Table IV.

TABLE IV. Air and apparent radiative surface temperatures on three summer days
(°C) [25].

	June 2–3 1960		*June* 29–30, 1961		*Aug.* 29–30, 1961	
	Max.	*Min.*	*Max.*	*Min.*	*Max.*	*Min.*
Screen air temperature (height of 1 meter)	23.4	8.1	25.4	12.0	29.7	13.7
Grass minimum temperature		2.8		6.7		11.1
Apparent surface temperature T^*						
(a) Short grass	30.5	4.0	35.9	6.6	31.7	10.2
(b) Long grass	25.1	4.7			32.4	10.2
(c) Evaporation tank	24.4	14.8			28.2	15.8
(d) Bare soil	38.9	5.1	44.1	8.7	39.0	11.7

Screen air temperature is much lower than apparent ground temperature during the daytime. The effect of different types of surfaces is also illustrated; for example, bare soil becomes much warmer than a grass or water surface. At night, screen temperatures are several degrees higher than apparent ground temperatures.

4.2. Subsurface Soil Temperatures

Thermocouples can be used to measure soil temperatures with an accuracy approaching 0.1°C. Figure 8 displays a typical summer diurnal cycle at O'Neill, Nebraska [26] under a grass surface. Air temperatures are included for comparison.

The diurnal wave of surface temperature is quickly damped with depth and has almost disappeared at 40 cm. There is a lag in times of maximum and minimum temperatures. At 2.5 cm the highest temperature occurs near 1530 CST, but at 20 cm the time is delayed to near 2030 CST. In contrast, air temperatures show only slight lag, from about 1330 CST at 10 cm to about 1430 at 40 cm.

Two vertical profiles of temperature obtained from Fig. 8 are displayed in Fig. 9. The large vertical temperature gradients that exist near the surface are illustrated; the impossibility of extrapolating profiles to obtain a surface temperature is evident.

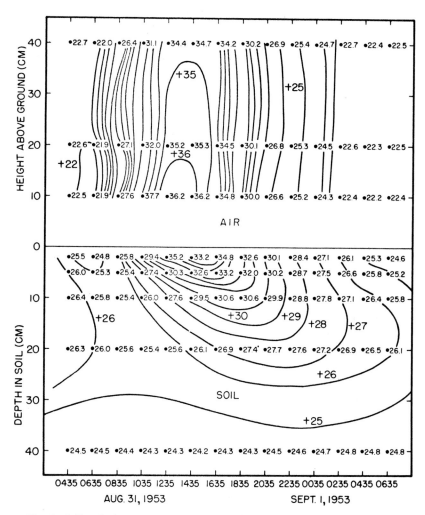

FIG. 8. Soil and air temperatures in °C at O'Neill, Nebraska, Aug. 31 to Sept. 1, 1953 for short grass surface [26].

The annual cycle of soil temperatures also shows a damping of amplitude and a phase lag with depth. The 1959 pattern for Ottawa (Experimental Farm), Canada is given in Fig. 10. The monthly value is the average of twice-daily readings at 0830 and 1600 EST. The surface was covered with snow in January, February, March, and

December. The isopleths are quite similar in shape to those of Fig. 8. The maximum is not reached until September; the annual amplitude is also greatly damped.

The temperature structure of snow, ice, and water will be considered in Chapters 15 and 18.

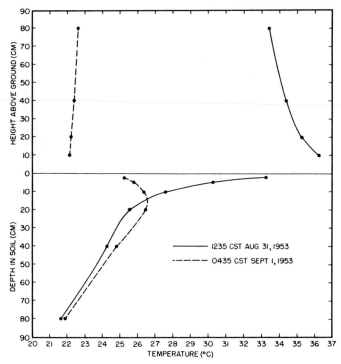

Fɪɢ. 9. Soil and air temperature profiles in °C at O'Neill, Nebraska, 1235 CST, Aug. 31 and 0435 CST, Sept. 1, 1953 [26].

4.3. Moisture in Bare Soil

Soil moisture is the source of water for evaporation. Changes of soil moisture with depth and with time must therefore be related in some way to surface water losses. Soil moisture is also an important regulator of soil temperature. Dry soil has a much lower specific heat than has water; hence, more heat is required to raise the temperature of wet than of dry ground an equivalent amount. For example, wet soils are slow to respond to the warmth of spring.

The *water table* is the upper surface of unconfined ground water. Above this there is a *capillary fringe*, where moisture is held against the pull of gravity by the forces of adhesion, cohesion, and surface tension. Capillary capacity is greatest when soil particles are small and the organic content is high.

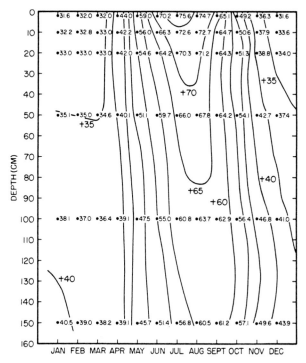

FIG. 10. Monthly average soil temperature isopleths in °F at Ottawa, Canada in 1959 (grass or snow surface).

Above the capillary fringe, the soil contains water in both liquid and vapor form. The movement of moisture in this region is not clearly understood, nor is the terminology standardized. Recent discussions of this problem have been given [27–30].

Field capacity is the amount of water remaining in well-drained soil when the velocity of downward flow into unsaturated soil has become small. It is expressed as a percentage of weight of oven-dry soil. Field capacity should be measured 1–2 days after the soil has become thoroughly wetted, the surface having been covered to reduce evaporation [27]. Soils are usually at field capacity in winter, but they acquire a moisture deficit in summer. *Percolation* is the downward motion of water through soil while *run-off* is the lateral movement of water toward streams and lakes. Although soil temperatures normally change only slowly with time, percolation can cause sudden rises or falls of as much as 5°C shortly after heavy rain has begun, e.g., Sarson [31].

Potential evaporation is the rate of evaporation from a moist surface under existing atmospheric conditions. Evaporation from bare ground proceeds in two stages. When the soil is at field capacity, evaporation is at the potential rate. When the soil becomes drier, the rate of evaporation depends not only upon atmospheric conditions but also upon the moisture flow rate through the soil. If E, E_p are the actual and potential evaporation rates, and if W, W_k are the actual and field capacity moistures, respectively, Budyko [9] suggests as an empirical approximation, when $W < W_k$:

(4.1) $$E/E_p = W/W_k$$

Equation (4.1) should be used with caution. A thin layer (not greater than 30 cm) of extremely dry surface soil inhibits upward movement of water from the moist soil below. Ideal meteorological conditions

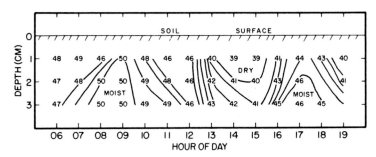

Fig. 11. Soil moisture content isopleths in bare soil at Tateno, Japan, Nov. 25, 1957. Moisture content is expressed as a percentage of water content per unit volume of soil [32].

for evaporation are brisk winds and very dry air, as in a foehn or chinook. However, if the top few centimeters of soil dry rapidly, the actual evaporation rate will be lower than under less favorable meteorological conditions.

An example of the diurnal cycle of soil moisture at Tateno, Japan is displayed in Fig. 11 [32]. The surface layers dry during the day when evaporation is greater than moisture flux from below.

It is difficult to measure soil moisture and no method is satisfactory. To obtain an absolute standard, soil samples are removed with an auger and are weighed before and after drying in an oven. However, this disturbs the soil and makes it impossible to take repeated samples at the same point. Slatyer and McIlroy [27] have discussed the advantages and disadvantages of other methods. Neutron scattering is considered to be the best currently available alternative. A source of quickly moving neutrons is introduced into the soil. Hydrogen atoms have a strong decelerating effect on neutrons; hence, a count of slowly moving neutrons is an index of the hydrogen content of the soil. Since water is the only significant source of hydrogen atoms in soil, the moisture content may be inferred. The method is not exact, the instrument cannot be used near the surface, and calibration is difficult.

Use is often made of a related parameter called *soil moisture tension*, the attraction of soil for water (dimensions of force per unit area and normally expressed in millibars). A *tensiometer* [26] is a porous cup buried in the soil and connected to a manometer by a water-filled tube. The instrument measures the osmotic pressure through the porous cup. It cannot be used in very dry soils or in freezing weather. Furthermore, the relation between soil tension and soil moisture depends upon the type of soil and upon the temperature. However, it is possible to obtain an approximate calibration for each location. The experimental determinations for O'Neill, Nebraska are given in Fig. 12 [26], which illustrates also the strong temperature dependence.

4.4. Evapotranspiration

Evapotranspiration is the combined water loss from a surface by evaporation and by transpiration from plants. *Potential evapotranspiration* is the evapotranspiration that would occur if soil moisture were not a limiting factor.

The growth of a plant depends upon the processes of photosynthesis and respiration. The general equations are as follows:

Photosynthesis (intermittent)
(4.2)

carbon dioxide + water + solar energy → carbohydrates + oxygen

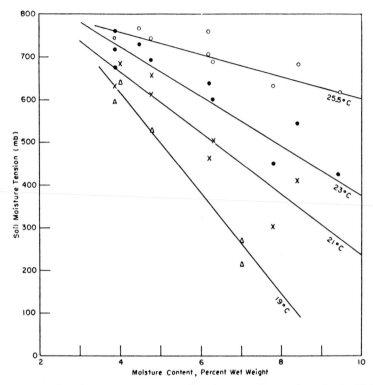

FIG. 12. Soil moisture content versus soil moisture tension at 10 cm depth, O'Neill, Nebraska for four temperatures [26].

Photosynthesis takes place through the action of chlorophyll and carotinoids; the solar energy of importance therefore depends upon the absorption spectra of these pigments (see ref. [33], Fig. 173, for spectra). On the average, less than 1% of incoming solar radiation Q_T is used in photosynthesis, although the value may reach 6 to 8% on a few occasions [34]. In addition to the fact that only certain wavelengths of light can be used, another reason for low photosynthetic efficiency is that a leaf becomes "light saturated" when the intensity is more than about $\frac{1}{10} - \frac{2}{10}$ of full noontime sunlight; the photosynthetic rate does not increase appreciably with larger values of Q_T [35].[2]

[2] J. D. Hesketh and D. N. Moss [*Crop Science* 3, 107–110 (1963)] have recently established that this is not universally true. Single leaves of sunflower, cane, and maize do not become light saturated up to at least full sunlight.

The water supply for the process comes from the plant roots through osmotic exchange with the soil.

Respiration (continuous)

carbohydrates + oxygen → carbon dioxide + water vapor +
(4.3)   combustion energy

Some of the carbohydrates produced by photosynthesis must be used to provide energy for plant growth. In this connection, equation (1.1) is not affected because the heat of combustion of carbohydrates is negligible in comparison with the other terms in the energy balance equation. Respiration of carbon dioxide is one-fifth to one-quarter of assimilation; a plant cover is therefore a sink for carbon dioxide during the day and a source at night. As might be expected from equation (4.2), photosynthesis can be increased at high light intensities by enriching the air with CO_2 [35].

Transpiration is the transfer of water to the atmosphere from within the plant. Ninety per cent of this takes place through stomata, small pores which average about 1–3% of total leaf area. Stomata tend to close at night or when there is drought or strong wind.

Water losses by transpiration are much greater than is actually required for plant growth. An oak tree may release as much as 150 gallons of water a day. This has been considered by some as an unavoidable evil. Plants require carbon dioxide, and when the stomata are open, outward diffusion of water vapor must occur. On the other hand, transpiration may be a physiological necessity, drawing up dissolved food from the roots.

Transpiration short-circuits normal channels for vertical soil moisture transfer. Water losses are therefore in general greater from a plant cover than from bare soil. In particular, the root depth is important. A plant's first response to water shortage is to extend its roots to deeper layers of soil. Transpiration therefore continues at the potential rate well below field capacity, ultimately falling off very abruptly to about one-tenth of what it could be [36]. The *wilting point* is the moisture content at which the soil cannot supply sufficient moisture to maintain the turgor of the plant. The relationship between wilting point and the limiting moisture for potential transpiration is not yet established [27]. It should also be noted that transpiration decreases when soil moisture becomes too great or in the presence of certain kinds of air pollution. Finally, Gardner and Ehlig

[37] have recently suggested that once plants wilt, an equation similar to (4.1) applies, i.e., the transpiration rate is roughly proportional to the available water content of the soil.

The way in which a plant reacts to a particular soil-atmosphere environment depends upon physiological as well as meteorological factors. For example, the micrometeorologist is not able to tell whether stomata are open or closed; he must work closely with the plant physiologist if many of the problems are to be solved.

4.5. The Lysimeter

Fortunately, an instrument is available for determining evapo-transpiration directly. Water losses from fields may be measured with the weighing lysimeter [38]. A large block of soil (up to 3 meters in depth and 6 meters in length) is lifted by derrick in the air; the pit is deepened and heavy-duty scales are placed in the hole; finally, the block of soil is lowered to its original position. An hydraulic method of weight recording can achieve an accuracy of ± 0.004 cm in water-depth equivalent.

Although the initial cost is high, and great care must be taken in site selection, the weighing lysimeter is an important micrometeorological research tool. It will yield estimates of the latent heat term Q_E with an accuracy of $\pm 10\%$. The installation at Davis, California [39] is a recommended model.

Other less expensive types of lysimeters have been designed but they do not always reproduce natural soil conditions. Reviews of lysimetry have been given [27, 33, 40, 41].

In addition to the discussion of Slatyer and McIlroy [27], useful references on the measurement of soil temperature and soil moisture are those of World Meteorological Organization [42, 43].

5. Soil Heat Transfer

5.1. Heat Transfer in a Solid

By convention downward heat flows into the soil are positive. The depth z increases positively downward from the interface $z = 0$.

Consider a homogeneous solid of thickness $(z_2 - z_1)$. Suppose that the upper and lower surfaces z_1 and z_2 are kept at temperatures T_1 and T_2 $(T_1 > T_2)$. Then it is well known that heat will flow by conduction from level z_1 to level z_2. The heat flux Q_G will depend upon the temperature difference, the thickness of the solid, and its conductivity.

For small $(z_2 - z_1)$ and $(T_1 - T_2)$, and for equilibrium or *steady state* conditions, i.e., temperatures not changing with time, the heat flux is given by equation (5.1):

$$(5.1) \qquad Q_G = k \frac{(T_1 - T_2)}{(z_2 - z_1)}$$

where k is by definition the thermal conductivity (cal/cm sec °C). Equation (5.1) may be written:

$$(5.2) \qquad Q_G = -k \frac{\partial T}{\partial z}$$

The conductivity of a material varies only slightly with temperature and may be considered as constant over the meteorological range of interest. However, there is a significant variation in various solids and liquids. Some typical conductivities are given in Table V.

Soil is not a true solid but consists of individual particles and aggregates in a medium of air, water, or ice. Particle diameters range from $1\,\mu$ (clay) to $100\,\mu$ (sand). Soil conductivity therefore depends on a number of factors:

1. The conductivity of the particles.
2. The size of the particles.

TABLE V. Conductivities of some substances
(cal/cm sec °C).

Water at 10°C	0.0014
Ice at −10°C	0.0055
Snow	
Density 0.1	0.00018
Density 0.5	0.0015
Density 0.9	0.0054
Very dry soil	0.0004–0.0008
Wet soil	0.003–0.008
Brick masonry	0.0015
Granite	0.0065
Iron	0.16
Still air	0.00006

3. The compaction of the mixture, i.e., the *porosity*, the ratio of the space between particles to the total space.
4. The soil moisture. Because water occupies different spaces in the wetting and drying cycles, the relationship between water amount and conductivity is not linear.

Soil conductivity is therefore variable in space and time. Although attempts have been made to measure it [26], the difficulties are such that equation (5.1) is not very useful for determining soil heat flux.

Because soil contains water and air, the possibility should not be overlooked of heat transfer by percolation, change of state, radiation, and convection. Theoretical models have been developed to predict the magnitudes of the effects but they all lack experimental verification.

5.2. The Fourier Heat Conduction Equation in One Dimension

Many flow problems in mathematical physics are formulated by a " box " model. Consider a small unit cube centered at the point (x, y, z) in space and extending out distances $(x \pm \Delta x)$, $(y \pm \Delta y)$ and $(z \pm \Delta z)$. The volume V of the cube is thus $8\Delta x\, \Delta y\, \Delta z$. If the mean temperature of the cube is T, then its rate of gain or loss of heat per unit time Δt will be

$$V \frac{\Delta(\rho c T)}{\Delta t}$$

where ρ is density and c is specific heat.

Suppose now that there are no heat flows in the x- and y-directions, but $(Q_G - \Delta Q_G)$ at the top of the box and $(Q_G + \Delta Q_G)$ at the bottom, per unit area. The net gain or loss of heat by the box is found by subtracting outflow from inflow and multiplying by the area.

$$(Q_G - \Delta Q_G)2\Delta x \, 2\Delta y - (Q_G + \Delta Q_G)2\Delta x \, 2\Delta y = -8\Delta Q_G \, \Delta x \, \Delta y$$

The two expressions must balance from which it follows that

(5.3)
$$\frac{\Delta Q_G}{\Delta z} = -\frac{\Delta(\rho c T)}{\Delta t}$$

Equation (5.3) may be written as

(5.4)
$$\frac{\partial Q_G}{\partial z} = -\frac{\partial(\rho c T)}{\partial t}$$

The product ρc is defined as *heat capacity*, and is nearly constant for many soils over times as great as a few days. Equation (5.4) may therefore be used to determine Q_G, given the value of heat capacity and a series of soil temperature profiles over a few hours. Integration of equation (5.4) from z_1 to z_2 yields

(5.5)
$$[Q_G]_{z_1}^{z_2} = -\rho c \int_{z_1}^{z_2} \frac{\partial T}{\partial t} \, dz$$

The right side of the equation may be evaluated graphically. Determination of Q_G from the left side depends upon a careful selection of a level of integration at which the heat flux is known to vanish. Referring back to Fig. 9, it will be seen that the nighttime soil temperature profile has a maximum between 10 and 20 cm. The heat flux at that level must therefore be zero, and the integration of equation (5.5) can be terminated at a depth of about 15 cm. When a series of profiles is examined, the maximum will be observed to move downward and become more difficult to locate. The significance of this is that while the upper layers are being cooled, the lower levels are still responding to daytime heating. The positions of heating and cooling maxima and minima can be followed and usually show a reasonable pattern (after slight smoothing) when plotted on a graph, permitting interpolation and some extrapolation.

The method provides an estimate of Q_G at the upper measurement level (2.5 cm in the case of Fig. 9). Since the soil is warming or cooling, there must be flux divergence between that level and the interface.

The correction is usually small, less than 10%, but for very precise determinations the soil temperature profile can be extended up to the interface by using the surface apparent radiative temperature given by equation (3.2).

In summary, the essential reason for determining Q_G from equation (5.5) rather than from (5.1) is that ρc may be estimated with more confidence than conductivity k.

Substitution of equation (5.2) in (5.4) with the assumption that ρc is constant leads to the Fourier heat conduction equation in one dimension:

$$(5.6) \qquad \frac{\partial T}{\partial t} = \frac{\partial}{\partial z}\left(\frac{k}{\rho c}\frac{\partial T}{\partial z}\right) = \frac{\partial}{\partial z}\left(K_G\frac{\partial T}{\partial z}\right)$$

The quantity, $K_G = k/\rho c$, is known as the *thermal diffusivity* and has the dimensions of cm²/sec. In Europe, use is also made of the *austausch coefficient* ρK_G with dimensions of gm/cm sec.

Assuming constant K_G, equation (5.6) may be solved for given initial and boundary conditions. If a periodic diurnal wave of surface temperature is imposed, for example, a damped wave descends downward with decreasing amplitude and increasing phase lag. The prediction agrees qualitatively with experimental results but there are discrepancies in detail and Lettau [44] concludes that K_G is not constant. This is not at all surprising in view of variations with depth in soil composition, compaction, and moisture.

5.3. Experimental Methods

The specific heat of dry soil c_d can be determined by standard laboratory techniques. It usually ranges from 0.16 to 0.21 cal/gm °C. The specific heat of water c_w is 1.0. Specific heat of moist soil c is given by

$$(5.7) \qquad c = c_w \cdot W + c_d(1 - W)$$

where W is the fraction by weight of the moisture in the soil. The quantity W may be determined by weighing a sample of soil before and after oven-drying.

The bulk density ρ is the density of a unit cube of moist soil. Provided that care is taken when removing a soil sample, bulk density may be determined by standard methods. Values are in the range of 1.0 to 1.6 gm/cm³ but are sometimes as high as 2 gm/cm³ for very compact soils.

One-dimensional heat flows may be measured directly in soils with a *heat-flow transducer* [26], a thin plate of known conductivity fitted with a thermopile. Difficulties arise because the thermal properties of the transducer may be significantly different from those of the soil and because the transducer acts as a barrier to the movement of soil moisture. The physical dimensions and thermal characteristics that reduce errors to a minimum have been discussed [45–47]. Commercially available instruments can now yield estimates of Q_G at the level of the transducer with an error of not more than 10%. There is, of course, also the problem of flux divergence, and the sensor should be located as close to the surface as is practicable, i.e., at a depth of 2–3 cm.

5.4. Some Estimates of Soil Heat Flux

Typical values of Q_G at Waltair, India [48] are given in Fig. 13. The dotted lines represent mean hourly values obtained from a heat-flow transducer. The solid lines were inferred from temperature measurements at 5, 15, and 30 cm, using basically the method of Lettau [44] for the solution of equation (5.6) with the assumption of constant K_G. The two curves are qualitatively similar, and illustrate the fact that soil acts as a heat reservoir, storing daytime energy for release at night.

For comparison, Fig. 14 displays the diurnal cycles at Chicago [49] based on the application of equation (5.5) to temperature measurements at depths of 1, 10, 20, 50, 100, 305, and 884 cm. In both Figs. 13 and 14, the heating part of the diurnal cycle is shorter but more intense than the cooling part. The maximum occurs about noon while the minimum is in the evening. At Chicago the cycle is greatly damped in winter.

Returning now to the energy balance equation (1.1) the terms Q_N and Q_G may be measured with reasonable accuracy (when care is taken) over soil. It is therefore possible to infer values of the sum $(Q_H + Q_E)$ that are within 10 or 15% of true values. One of the important problems in micrometeorology is the determination of the partition of energy between Q_H and Q_E, knowing their sum.

5.5. Soil Moisture Flux

A knowledge of soil moisture flux should in theory provide estimates of evaporation and of Q_E. This is not yet a feasible method

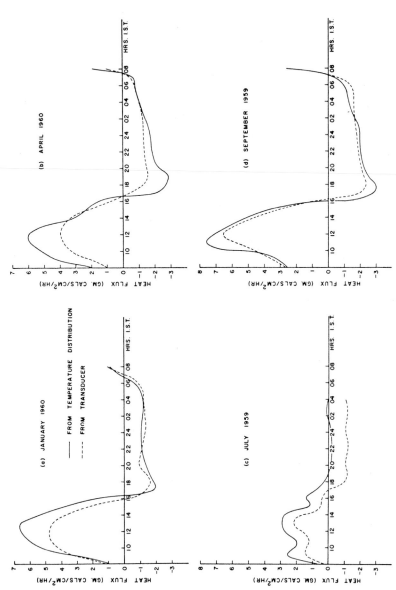

FIG. 13. Daily cycle of soil heat flux in four different months at Waltair, India; dashed line from transducer; solid line from temperature profiles [48].

because of the probable frequent occurrence of soil moisture flux divergence. Equations similar to (5.1) to (5.6) can be formulated, replacing temperature by water content and adding the effect of

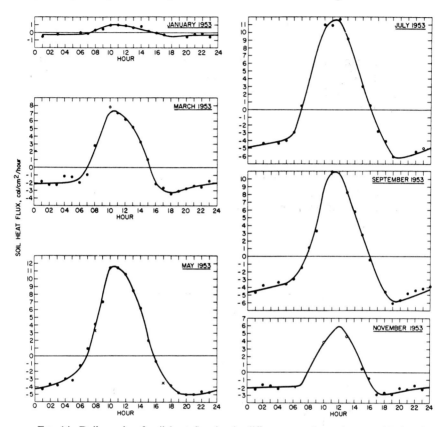

FIG. 14. Daily cycle of soil heat flux in six different months at Argonne National Laboratory, Chicago [49].

capillary transfer. For example, Philip (30) has obtained solutions for the equation

$$(5.8) \qquad \frac{\partial W}{\partial t} = \frac{\partial}{\partial z}\left(K_s\frac{\partial W}{\partial z}\right) + \frac{\partial C}{\partial z}$$

where W is the volumetric soil water content, K_s is soil moisture diffusivity, and C is the soil moisture capillary conductivity. Soil temperatures are assumed to be isothermal. There are difficulties in experimental verification and the model is still being tested.

Maruyama [32] begins his analysis with equation (5.9)

$$(5.9) \qquad \frac{\partial}{\partial t}(W + W') = \frac{\partial}{\partial z}\left(K_S \frac{\partial W}{\partial z} + K_S' \frac{\partial W'}{\partial z}\right)$$

where the prime denotes soil moisture in vapor form. Neglecting the case of very dry soil, Maruyama believes that W' and K' are small in comparison with W and K_S, respectively. Thus, equation (5.9) may be simplified to the form:

$$(5.10) \qquad \frac{\partial W}{\partial t} = \frac{\partial}{\partial z}\left(K_S \frac{\partial W}{\partial z}\right)$$

Measurements of soil moisture were made at depths of 1.2 and 3 cm, from which the moisture diffusivity was estimated in equation (5.10). Finally, evaporation rate was computed from an equation similar to (5.2):

$$(5.11) \qquad E = K_S \frac{\partial W}{\partial z}$$

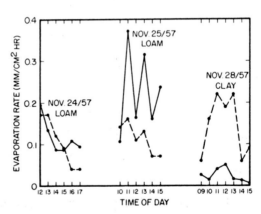

FIG. 15. Evaporation rates at Tateno, Japan calculated from soil moisture profiles (solid lines) (see Fig. 11) and from a small weighing lysimeter (dashed lines) [32].

The resulting estimates of evaporation for three days at Tateno, Japan are given by the solid lines in Fig. 15. Included also are the values obtained from a small weighing lysimeter, 20 cm in depth and

25 cm in length. The agreement is not good but this may be due in part to lysimeter errors. In addition, the soil moisture values may not be reliable.

In conclusion, the search for methods of determining evaporation through soil moisture flux is to be encouraged. The large weighing lysimeters are costly and not portable; an accurate and relatively simple approach using soil moisture profiles would therefore be a major step forward.

6. Air Temperature and Humidity near the Earth's Surface

6.1. Factors Influencing Air Temperatures

The surface boundary layer responds much more quickly to energy changes at the interface than does the soil. Referring back to Fig. 8, the lag in time of maximum temperature was only about half an hour between 10 and 40 cm in the air but was 3 hr between 10 and 20 cm in the ground. Conduction is the principal method of heat transfer in the soil but its contribution is negligible in the atmosphere.

The factors that determine air temperatures are radiative flux divergence, *advection* (horizontal motion of the wind), *convection* (vertical motion of the air), and latent heat exchanges by evaporation and condensation. In the absence of fog, low cloud, or precipitation, the latent heat process occurs only at the ground.

Advection is important when the surface of the earth is not uniform. Examples are the motion of warm air across a cooler lake, and the flow of air from prairie to forest. Further discussion will be postponed until Chapter 12 but it may be stated now that the principal effect of advection is to change the convective characteristics of the air. An advective term does not appear in the energy balance equation (1.1) because the wind flow is usually horizontal and cannot contribute directly to heat exchanges in a perpendicular direction.

There are three types of convection—forced, free, and natural. *Forced convection* occurs when air blows across a rough surface. Turbulence develops (see Chapters 7 and 8) and the air from different levels is mixed, smoothing away large temperature differences. As the wind increases or as the ground becomes rougher, forced convection becomes more vigorous.

Free convection is caused by density or buoyancy differences within a moving fluid. It develops, for example, as a result of uneven heating of the ground. The air over "hot spots" tends to rise, carrying with it heat, moisture, and pollution, sometimes to heights of several kilometers. *Natural convection* occurs over a heated surface when there is no wind. It has been studied extensively in the laboratory.

Air is compressible, which implies that atmospheric density decreases with height. It is well known that adiabatic expansion of air causes cooling while adiabatic compression (as in a bicycle pump) causes warming. A rising bubble of air therefore moves through levels of decreasing density, expands, and cools. In the hypothetical case of no mixing or heat exchange between the rising air and its environment, cooling is said to be at the *dry adiabatic rate* (unsaturated air) or the moist *adiabatic rate* (saturated air in which the latent heat of condensation must reduce the rate of cooling). Rising air is therefore subjected to changing buoyancy forces, dependent upon the density or temperature stratification of the environment. For example, if a heated bubble of air moves upward through a region in which temperature is steady with height, the bubble will expand and cool until ultimately it is at a lower temperature than adjacent air. Negative buoyancy will therefore develop.

Lapse rate is the rate of temperature decrease with height in the atmosphere. The various classes of lapse rate are the *inversion* (negative lapse rate), *isothermal* (zero lapse rate), *weak lapse*, *dry adiabatic* (0.98°C/100 meters or 5.4°F/1000 ft), and *superadiabatic* (greater than 0.98°C/100 meters). Another frequently used term is vertical *temperature gradient*, which is lapse rate with reversed sign. For example, an inversion is a negative lapse rate but a positive temperature gradient.

Lapse rate has been used classically as an index of vertical stability. An atmosphere in which temperature decrease with height is greater than, equal to, or less than the adiabatic rate is said to be in *unstable*, *neutral*, or *stable* equilibrium, respectively.

Although an oversimplification, this view of vertical stability is useful. For example, *inversion* has become a household word wherever atmospheric pollution is a problem. However, since forced convection depends also on wind speed and ground roughness, a true index of vertical mixing must include wind as well as lapse rate (see the Richardson number in Chapter 9).

In some applications, actual temperature is replaced by *potential temperature*, the temperature a parcel of dry air would have if brought adiabatically from its initial state to a standard sea-level pressure of 1000 mb. In the surface boundary layer, the relation between the two

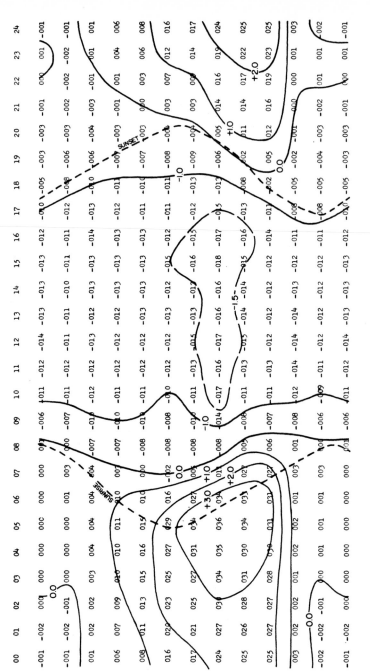

FIG. 16. Average temperature differences (°F) by hours of the day and by months between the 20- and 300-ft levels obtained from a television tower near Detroit for the period of Aug. 1955 to Aug. 1962.

is given with sufficient accuracy by equation (6.1):

(6.1) $$\theta = T + \Gamma z$$

where θ, T are the potential and actual temperatures, respectively, and Γ is the dry adiabatic lapse rate. When some writers use the term "inversion," they are referring to an inversion of potential temperature.

6.2. Diurnal and Annual Patterns of Air Temperature Differences

Figures 16 and 17 display for each hour of the day mean monthly vertical temperature differences for Detroit, Michigan and Resolute in the Canadian Arctic. The Detroit measurements are from the 20- and 300-ft levels of a television tower; the Resolute observations come from the 6.4- and 100-ft levels of a micrometeorological tower.

It can be seen that temperature decreases with height during daylight hours while at night there is usually an inversion. The long period of polar darkness at Resolute results in a greater frequency of inversions than at Detroit. These results are of course true only on the average. On any specific occasion, the macro- and mesoscale features exert a major influence.

In general, a superadiabatic lapse rate occurs on sunny days or when cold air is flowing across a warmer surface. The adiabatic lapse rate, on the other hand, is found infrequently. It will occur in transition periods lasting a few minutes separating inversion from superadiabatic conditions, shortly after sunrise and near sunset, for example. It is also observed when the air is well stirred by strong winds and rough ground.

Inversions are classified in two ways: (a) according to the method of formation, and (b) according to height of base, thickness, and intensity.

The *radiation inversion* is common at night, particularly under clear skies with light winds. A typical radiation inversion profile is given in Fig. 18, obtained with a tethered balloon at Paisley, Canada; a daytime profile is included for comparison.

The base of the radiation inversion is surprisingly not always at the surface of the earth but is often at a height of from 2 to 5 cm, a fact of some practical importance in frost protection. Romanova and Kaulin [50] reported one case in which the grass minimum temperature was 7.7°C warmer than the temperature at a height of 2 cm.

Hour (C.S.T.)	00	01	02	03	04	05	06	07	08	09	10	11	12	13	14	15	16	17	18	19	20	21	22	23
JANUARY	012	012	012	012	012	012	011	011	011	010	011	010	010	011	011	010	010	010	011	011	011	011	012	012
FEBRUARY	009	009	009	010	009	009	009	010	010	009	009	009	008	008	008	009	010	011	010	010	010	010	010	009
MARCH	011	011	011	011	012	012	012	012	012	012	011	007	004	004	001	001	006	009	009	011	010	010	010	011
APRIL	014	014	015	013	013	014	014	011	005	006	004	001	-001	-001	-002	-001	002	002	003	-006	012	013	013	014
MAY	006	006	006	005	003	002	000	-001	000	-002	-002	-004	-004	-004	000	-002	-002	-003	000	002	004	005	005	005
JUNE	001	001	001	000	-001	-002	-004	-005	-006	-007	-008	-008	-009	-007	-006	-006	-007	-005	-003	-001	001	000	000	001
JULY	002	002	002	001	001	-002	-003	-004	-006	-007	-008	-008	-009	-009	-007	-007	-007	-004	-003	-002	000	001	000	001
AUGUST	004	005	005	003	002	002	000	-002	-003	-004	-005	-006	-008	-007	-006	-005	-005	-004	-002	000	002	002	003	004
SEPTEMBER	002	002	002	002	002	002	001	001	000	-002	-003	-004	-005	-005	-004	-004	-003	-002	-001	000	001	001	001	002
OCTOBER	005	004	005	005	004	005	004	004	005	004	003	004	002	003	004	005	005	005	005	005	005	005	005	005
NOVEMBER	011	011	011	011	011	011	010	010	010	011	010	011	011	011	011	011	010	010	011	011	010	011	011	011
DECEMBER	011	011	011	011	011	010	010	010	010	011	011	011	011	011	011	012	012	012	012	011	012	012	011	011

Fig. 17. Average temperature differences (°C) by hours of the day and by months between the 6.4- and 100-ft levels at Resolute in the Canadian Arctic for the period of Aug. 1957 to Feb. 1961.

The morning breakup of the inversion is usually quite rapid. A superadiabatic lapse rate is established near the ground, and convection quickly erodes the inversion from below. In atmospheric pollution studies, the process has been called a *fumigation* [51]. Stack gases are often trapped in a stable layer during the night. When the

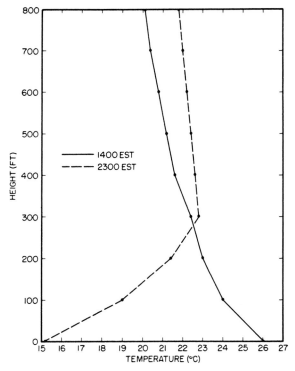

FIG. 18. Vertical temperature profiles at Paisley, Canada, July 19, 1962; ordinate is height above balloon release level (about 5 ft).

morning turbulence reaches plume height, pollution is carried down to ground level by vigorous mixing in relatively high concentrations. The fumigation lasts about 15 min.

An *advection inversion* forms when warm air blows across a cooler surface. The cooling of the air may be sufficient to produce fog. When an inversion over lake or ocean in summer moves inland, a continuous fumigation may occur during the daytime. Figure 19 displays three typical June temperature profiles from Douglas Point, located on the east shore of Lake Huron. There is no inversion in the lowest 200 to 300 ft because of strong solar heating of the ground. With increasing distance inland from the lake, the base of the inversion rises.

The *subsidence inversion* is associated with slowly descending air aloft in high pressure areas, with consequent adiabatic warming.

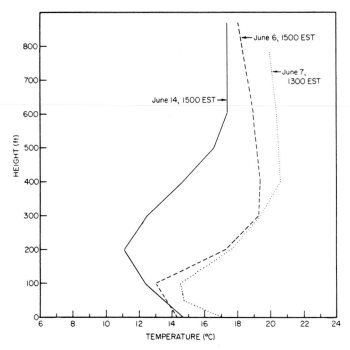

FIG. 19. Vertical temperature profiles at Douglas Point, Canada on three typical lake-breeze days; ordinate is height above balloon release level (about 5 ft).

The most widely studied example occurs in California on the eastern edge of a semipermanent high pressure area. The inversion may at times be a kilometer thick and can persist for weeks.

An *evaporation inversion* may occur after a summer shower or over an irrigated field. Whenever Q_N is positive but less than $(Q_G + Q_E)$, Q_H must be negative with a resulting temperature inversion.

6.3. Precipitation and Fog

Micrometeorologists have shown some interest in fog [52, 53], but rainy periods have been considered as meso- and macroscale phenomena. Precipitation usually originates well above the boundary layer and is associated with scales of motion that are not properly included in micrometeorology. Another branch of meteorology, called *cloud physics*, is concerned with these phenomena.

6.4. Humidity near the Earth's Surface

Water vapor exerts a pressure e which is only 1 to 2% of the pressure p of the atmosphere. Nevertheless, water vapor is an important gas because of its latent heat properties. A useful index of humidity is *mixing ratio* X, the ratio of the mass of water vapor to that of dry air in a given volume of moist air. It can be shown that

$$(6.2) \qquad X = 0.622e/(p - e) \simeq 0.622e/p$$

Mixing ratio is independent of temperature when there is no condensation or evaporation.

Another measure of the water vapor content of the air is *specific humidity* q, the ratio of the mass of water vapor to the total mass of air. Its value is given by equation (6.3):

$$(6.3) \qquad q = 0.622e/(p - 0.378e) \simeq 0.622e/p \simeq X$$

The diurnal variation of mixing ratio is given in Fig. 20 for 11 levels on a 100-ft tower at Suffield, Canada averaged over the month of March 1946; the corresponding temperatures are given for comparison in Fig. 21. The mixing ratio is at a minimum in the afternoon because of moisture flux divergence; water vapor is being transferred upward more quickly than water is being evaporated from the ground. The reverse situation occurs in the evening. The corresponding flux divergence of heat may be inferred from Fig. 21.

6.5. The Measurement of Temperature and Mixing Ratio

Climatic information on temperature and mixing ratio is obtained from mercury "dry-bulb" and "wet-bulb" thermometers in Stevenson screens. The mixing ratio can be determined from standard tables, given the readings from a pair of mercury thermometers, one of which has its bulb covered with a moistened wick.

Important considerations in micrometeorological studies are ventilation, shielding (from radiation), and speed of response. When the sensor is located within a few centimeters of the ground, physical

TIME OF DAY (MST)

Height (Ft)	00	01	02	03	04	05	06	07	08	09	10	11	12	13	14	15	16	17	18	19	20	21	22	23	24
100	3.4	3.4	3.3	3.4	3.4	3.3	3.4	3.4	3.3	3.2	3.2	3.2	3.1	3.0	3.0	3.0	3.0	3.1	3.3	3.4	3.5	3.5	3.4	3.4	3.4
90	3.4	3.4	3.4	3.4	3.4	3.4	3.4	3.4	3.3	3.2	3.2	3.2	3.2	3.0	3.0	3.0	3.1	3.1	3.3	3.6	3.5	3.5	3.4	3.4	3.4
80	3.4	3.4	3.4	3.4	3.4	3.4	3.4	3.4	3.3	3.2	3.2	3.2	3.2	3.0	3.0	3.0	3.0	3.2	3.3	3.5	3.5	3.5	3.5	3.4	3.4
70	3.4	3.4	3.4	3.4	3.4	3.4	3.4	3.4	3.4	3.2	3.2	3.2	3.0	3.0	3.0	3.0	3.0	3.2	3.3	3.5	3.6	3.5	3.5	3.5	3.4
60	3.4	3.4	3.3	3.4	3.4	3.4	3.4	3.3	3.4	3.3	3.2	3.3	3.2	3.1	3.0	3.0	3.0	3.2	3.3	3.6	3.6	3.6	3.5	3.5	3.4
50	3.4	3.4	3.4	3.4	3.4	3.4	3.4	3.4	3.4	3.3	3.2	3.3	3.2	3.1	3.0	3.0	3.1	3.2	3.4	3.6	3.6	3.6	3.5	3.5	3.4
40	3.4	3.4	3.4	3.4	3.5	3.4	3.4	3.4	3.4	3.3	3.2	3.3	3.2	3.1	3.0	3.0	3.1	3.2	3.4	3.6	3.6	3.6	3.6	3.6	3.4
30	3.4	3.4	3.4	3.4	3.5	3.4	3.4	3.4	3.4	3.3	3.3	3.3	3.3	3.2	3.1	3.1	3.2	3.2	3.5	3.7	3.6	3.6	3.6	3.6	3.5
20	3.5	3.5	3.5	3.5	3.5	3.4	3.5	3.4	3.5	3.4	3.3	3.4	3.4	3.2	3.2	3.2	3.2	3.3	3.6	3.7	3.7	3.7	3.6	3.6	3.5
10	3.5	3.5	3.5	3.5	3.5	3.4	3.5	3.4	3.5	3.4	3.2	3.5	3.5	3.3	3.2	3.2	3.4	3.4	3.6	3.7	3.8	3.7	3.6	3.6	3.5
5	3.6	3.5	3.5	3.5	3.5	3.4	3.5	3.4	3.4	3.4	3.4	3.5	3.4	3.3	3.2	3.2	3.3	3.4	3.6	3.7	3.8	3.7	3.7	3.7	3.6

FIG. 20. Mean hourly mixing ratios (gm/kg) for March 1946 at Suffield, Canada.

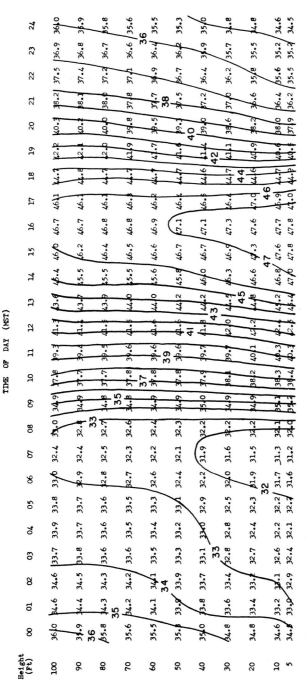

FIG. 21. Mean hourly temperatures (°F) for March 1946 at Suffield, Canada.

dimensions become important and great care must be taken not to disturb the natural wind.

The response of a thermometer is given by its *time constant*, the time required for the instrument to respond to 63.2% $(1 - 1/e)$ of a discrete change in environment temperature for given flow rate of air. A standard mercury thermometer has a time constant of more than a minute while the value may be much less than a second in the case of some thermocouples. The "best" sensor depends upon the purpose of the investigation. In order to obtain meaningful vertical profiles of temperature, a time constant of at least 1 min is desirable; on the other hand, if the rapid turbulent fluctuations are being studied (see Chapter 9), time constants of less than a second are often used.

Temperatures should be measured with an accuracy of 0.1°C. This can be done with thermocouples, resistance thermometers, or thermistors. A fine-wire thermocouple has the advantage that it may be exposed at night without shielding or ventilation, provided that the natural wind is not calm. The resistance thermometer is recommended for towers designed to operate with little maintenance. Thermistors have had a bad reputation for calibration drift but better models are becoming available commercially.

Profiles of mixing ratio may be obtained by using pairs of sensors, one of which is wrapped in a moistened wick. The moisture supply is difficult to regulate, and the instrument cannot be left unattended for more than a few hours. Furthermore, when the weather is very cold, meaningful observations are not usually possible.

Other techniques that have been used include the lithium chloride method, the dew-point or frost-point hygrometer, the refractometer, and spectrophotometry. None is completely satisfactory.

7. Wind Flow over Homogeneous Surfaces

7.1. The Essential Problem

Since heat and water vapor fluxes are dependent upon wind and turbulence, it is necessary to consider these latter factors before discussing Q_H (Chapter 9) and Q_E (Chapter 10).

Physical models for the flow of air in the boundary layer are based on the assumption that the surface of the earth is homogeneous and uniformly rough; furthermore, a geostrophic wind is supposed to have been blowing for a sufficiently long time for equilibrium conditions to exist. These rather implausible assumptions developed historically from the fact that this was the simplest kind of model that could be specified mathematically. Furthermore, the predictions could be tested in the controlled laboratory conditions of the wind tunnel.

Air flow over the ground was likened to the behavior of a fluid moving over a flat plate. However, as recently pointed out by Lettau [54], the analogy is poor because the flat-plate boundary layer thickens with increasing distance from its leading edge whereas the atmospheric boundary layer is presumably of constant depth over the entire "infinite plane." Lettau suggests that it would be more relevant to use pipe flow; although there is no restricting upper surface in the atmosphere, there is at least supposed to be a steady geostrophic wind corresponding to the flow at the center of a pipe.

The requirement for an "infinite plane" leads to a situation in which the micrometeorologist is never certain whether his models are incomplete or whether his experimental conditions are inadequate. Reproducibility of results from different parts of the world is one method of verification. However, there is a natural human tendency for data that confirm established models to appear more frequently in the published literature than "anomalies," which may be attributed to experimental error. The essential problem is that it is almost impossible to undertake a *controlled* atmospheric experiment in the laboratory sense. These remarks are intended to caution but not to discourage the reader.

7.2. Dimensional Analysis and Similarity Theory

It is an observed fact that the fundamental physical laws are dimensionally homogeneous (in mass, length and time, or combinations thereof). Consider as a trivial example, the equation

$$y = ax + b$$

If x and y are lengths, b must be a length while a is dimensionless; this, of course, cannot be proved formally.

The assumption can be used to obtain partial information about the nature of the relation existing among any given set of physical variables. Buckingham's theorem [55] states:

1. If an equation is dimensionally homogeneous, it can be reduced to a relationship among a complete set of dimensionless products.
2. The number of products in a complete set of dimensionless products of the variables $x_1 x_2 \ldots x_n$ is $(n - r)$, where r is the rank of the dimensional matrix of the variables.

The theorem may be explained by an example. A periodic temperature $(T = T_0 \cos pt)$ is applied at the surface of a semi-infinite homogeneous solid with thermal diffusivity K_G. Isothermal conditions $(T = T_0)$ exist at time $t = 0$. The dimensional analysis is begun by determining through inspection or physical intuition the relevant parameters. In the present case, if it be assumed that these be T_0, T, z, K_G, p, then the physical dimensions of the parameters are written in matrix form shown here.

	T_0	T	z	K_G	p
Length	0	0	1	2	0
Temperature	1	1	0	0	0
Time	0	0	0	−1	−1

The rank of the matrix is 3, while the number of parameters is 5; there is therefore a relationship between two dimensionless products. One of these products must be T/T_0, since combinations of z, K_G, and p cannot be made dimensionless if one of T or T_0 is included with them. By inspection, the other product is $z(p/K_G)^{1/2}$, or some related combination such as $z^2 p/K_G$.

The required equation is therefore of the form: $T/T_0 = F[z(p/K_G)^{1/2}]$ where F is an unknown function. The form of F may be determined from experimental data. However, an assessment of the usefulness of dimensional reasoning can be made here by comparison with the complete solution obtained by other methods:

$$T/T_0 = \exp\{-z(p/K_G)^{1/2}\}$$

The method is of particular importance when an experiment cannot be controlled by keeping all but two of a large number of variables constant. The data are difficult to comprehend when in standard form. However, if six variables, for example, may be reduced to two groups of dimensionless parameters, the data may be plotted on a single sheet of ordinary graph paper. Also, the results will be independent of the units of measurement.

It is emphasized that a dimensional analysis suggests but never proves the existence of a relation; the results must always be tested experimentally. There is a formal way of doing the analysis [55] and the method is not difficult. However, a large degree of physical intuition is required in the initial selection of pertinent variables.

For fluid flow through a pipe, the parameters of importance are u, a characteristic velocity such as the average flow rate in the pipe z, a characteristic thickness such as pipe diameter, and v, the *kinematic viscosity* of the fluid (in units of cm^2/sec). The dimensionless combination is the *Reynolds number*, Re:

(7.1) $Re = uz/v$

The Reynolds number is a fundamental tool in fluid mechanics and has been particularly useful in defining the transition from *laminar* (fluid moving smoothly in parallel streamlines) to turbulent flow. However, because of the existence of buoyancy forces, the Reynolds number is of little use in micrometeorology.

Similarity theory is essentially a restatement of the cause and effect relation. If conditions surrounding two experiments are identical, the results should be similar. In many micrometeorological cases, however, very little control over experimental conditions is possible. For example, the roughness of the ground will be different at various field stations where measurements may be made. Fortunately, Buckingham's theorem is applicable in most cases. It is not necessary to keep all the variables constant at two sites or at the same site on two different occasions. Complete similarity is achieved if any independent set of dimensionless products of the variables remains constant. For example, a number of pipes with different diameters have similar flows if the Reynolds numbers are all the same.

Similarity theory predicts universal functions which may be determined experimentally and tested for reproducibility at other locations. Similarity predictions in the atmosphere have been largely verified with the exception of relations that involve the cross-wind direction, e.g., the lateral component of turbulence. The disturbing factors in this latter case appear to be mesometeorological in origin.

Dimensional analysis and similarity theory are used extensively in micrometeorology. It is no exaggeration to state that practically the whole theory of the surface boundary layer is derived from these two fundamental principles.

7.3. Viscosity and Shearing Stress

It will be assumed in this section that an average wind $\bar{u}$ can be specified. The bar denotes a mean value at any point in space. The way in which the average is obtained and a discussion of its turbulent fluctuations will be given in Chapter 8.

Suppose that a fluid is flowing over a horizontal surface. If $\bar{u}$ is constant with height, the fluid is slipping over the surface without restraint and is said to be *nonviscous*. The atmosphere is *viscous*, however, which means that the ground exerts frictional drag, and the wind speed increases with height. The force of retardation per unit horizontal area is known as the *surface shearing stress* τ_0 in units of dynes/cm².

Shearing stress originates at the ground but exists through the surface boundary layer. *Wind shear* is defined as a change of wind velocity with height (or in the horizontal), and is a consequence of shearing stress. The magnitude of shearing stress depends upon the geostrophic wind speed, the surface roughness, and buoyancy.

An empirical index of friction is the *drag coefficient* C_D (nondimensional) defined by the equation

(7.2) $$\tau_0 = C_D \rho \bar{u}^2$$

where $\bar{u}$ is the wind speed at some convenient reference level such as 10 meters.

For laminar flow in which shearing stresses are caused entirely by molecular agitation, it may be inferred from the kinetic theory of gases (e.g., Sutton [56]), that there is a linear relation between shearing stress and wind speed gradient:

(7.3)
$$\tau = \mu \frac{\partial \bar{u}}{\partial z}$$

where the coefficient μ is defined as *dynamic viscosity* and is related to kinematic viscosity v, by the equation

(7.4)
$$v = \mu/\rho$$

The two viscosities are given in standard tables and vary only slightly with temperature in the surface boundary layer. For a temperature of 20°C, μ and v for air are 1.81×10^{-4} gm/cm sec and 0.150 cm²/sec, respectively.

The quantity $(\tau/\rho)^{1/2}$ has the dimensions of a velocity and it is convenient to define a *friction velocity* u_*:

(7.5)
$$\frac{\tau}{\rho} = u_*^2 = v \frac{\partial \bar{u}}{\partial z}$$

The kinematic viscosity v has the same dimensions as the coefficient K_G of equation (5.6); thus v is analogous to a diffusivity. What is being transferred vertically? It was emphasized in earlier sections that temperature gradients result in heat flows. Similarly, wind shear transfers momentum from regions of strong winds to regions of light winds, i.e., downward. Molecules are moving randomly in all directions; those that move up decrease the average momentum of the air at the higher level; molecules moving down, on the other hand, increase the momentum of the lower layers. Although momentum is being transferred in laminar shear flow, it is important to remember that there is no net mass exchange of air in the vertical direction.

For turbulent flow, molecular agitation is trivial in comparison with the bulk motion of eddies, bubbles, or parcels of air. Momentum is still being transferred downward but at a much more rapid rate. Equation (7.5) is generalized by introducing K_m, the *momentum eddy diffusivity*:

(7.6)
$$\frac{\tau}{\rho} = u_*^2 = (K_m + v) \frac{\partial \bar{u}}{\partial z} \simeq K_m \frac{\partial \bar{u}}{\partial z}$$

The coefficient K_m is defined by equation (7.6). However, the analogy with equation (7.5) should not be pressed too far. Whereas viscosity is constant for a given fluid at fixed temperature and pressure, eddy diffusivity varies with height above ground, the surface roughness, and buoyancy. One may therefore wonder about its physical significance. Another difficulty is that the classical momentum transfer theory is too simplified. Lettau [57] has recently noted that when a turbulent fluid element moves upward, it is not necessarily replaced by air from the upper level; instead, some lateral motion is to be expected because of the three-dimensional character of the flow.

An assumption of great importance in the study of boundary layer flows is that shearing stress and friction velocity are approximately constant with height, provided that the underlying surface is homogeneous, the pressure gradient is constant, and the Coriolis force may be neglected. There are a number of theoretical justifications (e.g., Calder [58]), and much speculation about the depth of the layer in which constancy may be assumed with sufficient accuracy. Monin and Obukhov [59] suggest an average value of 50 meters but during very strong inversions the vertical gradient of shearing stress may become important at a height of 2 meters [60]. For practical purposes, the surface boundary layer is often defined as that layer in which the shearing stress does not vary by more than 5%. Using this criterion [26], it was estimated that the thickness of the surface layer at O'Neill, Nebraska during the Great Plains program was about 8 meters at night and more than 16 meters during the day. Furthermore, the thickness of the surface layer increased with increasing geostrophic wind during daylight hours; there was no definite trend at night.

Another important parameter is the *rate of viscous dissipation* ε per unit mass of air (cm^2/sec^3), i.e., the work done per unit time per unit mass in converting turbulent energy into heat. Although viscous dissipation proceeds more rapidly in a shear zone, it occurs whenever there is turbulence.

Energy dissipation converts turbulent motion into frictional heat. However, the rate of heating is negligible in comparison with the other terms in equation (1.1). Brunt [61] estimated that the heat of viscous dissipation in the lowest kilometer of the atmosphere is about 1% of net solar radiation when averaged over the entire globe.

When there are no buoyancy forces, the parameters that determine the rate of energy dissipation are the height above the surface and a reference velocity for the flow, taken for convenience to be the friction velocity: u_* (with the additional assumption that u_* is constant with height). A dimensional argument leads to

$$(7.7) \qquad k\varepsilon = u_*^3/z$$

where k is a dimensionless constant, known as *von Kármán's constant*.

7.4. The Vertical Wind Profile in the Absence of Buoyancy

Taylor [62] has noted that the production of shear-flow energy (transfer of energy from the mean flow to turbulence) is given by equation (7.8) when buoyancy forces may be neglected:

$$(7.8) \qquad P = \frac{\partial}{\partial z}(\bar{u}\, u_*^2)$$

Since energy production and energy dissipation must balance, equations (7.7) and (7.8) may be used to yield (assuming constant u_*):

$$(7.9) \qquad \frac{\partial \bar{u}}{\partial z} = u_*/kz$$

Also, equation (7.6) and (7.9) may be combined to show that

$$(7.10) \qquad K_m = k u_* z$$

Equation (7.9) has been derived in various ways but the method used here [63] appears to be the most direct. Alternatively, equation (7.9) may be considered simply as an empirically established result of wind tunnel boundary layer flow.

If $\bar{u} = 0$ at $z = z_0$, equation (7.9) may be integrated:

$$(7.11) \qquad k\bar{u}/u_* = \ln z/z_0$$

The quantity z_0 is a constant of integration known as the *roughness length*.

A table of values for z_0 has been given by Sutton [56]. The quantity varies from a value of 0.001 cm (mud flats and ice) to 9 cm (thick grass 50 cm high). When the surface is very rough as in the case of forest, an additional parameter is introduced, the *zero-plane displacement d*. Equation (7.11) is rewritten as

$$(7.12) \qquad k\bar{u}/u_* = \ln (z-d)/z_0$$

The modification is empirical, providing a better fit for experimental data. In either case the equations should not be used for heights of the same order of magnitude as z_0 or d.

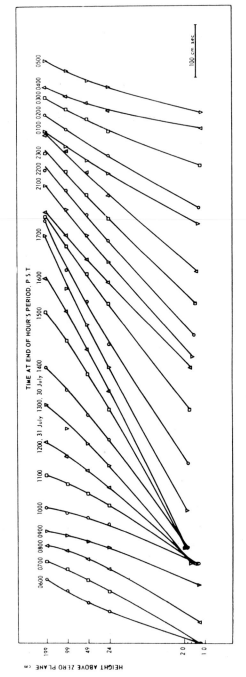

FIG. 22. Wind profiles at Davis, California July 30-31, 1962 [64].

Von Kármán's constant is often assumed to have a value of 0.41 in the wind tunnel but Lettau [54] has suggested a revision to 0.428 as a result of his re-examination of early data of flow in rough pipes. In the atmosphere, the logarithmic profile has been found to hold in adiabatic conditions over a semi-infinite plane, and von Kármán's constant has a value close to 0.4. However, experimental data are not sufficiently precise to distinguish between 0.41 and 0.428. Finally, Lettau [57] has recently suggested a physical-statistical interpretation of von Kármán's constant. It may be an index of the *three*-dimensional nature of the vertical transfer of fluid elements; the simple *two*-dimensional explanation of momentum transfer given in Section 7.3 might lead to a value of unity for k.

7.5. The Vertical Wind Profile in a Nonadiabatic Atmosphere

Although the logarithmic wind profile is important in wind tunnel studies of shear flow, it is not particularly useful in the atmosphere. The lapse rate is rarely adiabatic and buoyancy forces must be considered.

Typical wind profiles are given in Fig. 22 [64]. The departures from straight lines are not large but nevertheless it can be seen that the curvatures are different at night than during the afternoon.

An examination of similar data led to the empirical proposal by Deacon [65] that the wind profile be specified by its curvature:

$$(7.13) \qquad\qquad \frac{\partial \bar{u}}{\partial z} \propto z^{-\beta}$$

where β is greater than, equal to, and less than unity for super-adiabatic, adiabatic, and stable lapse rates, respectively. When the assumption was made that β was constant with height, the proportionality factor could be determined by comparison with the adiabatic case. Nevertheless, equation (7.13) is empirical and with the introduction of the Monin-Obukhov model (see below), it appeared that Deacon's equation might become merely of historical interest. However, the curvature of the wind profile is the basis for new studies by Lettau [66]. Rider [67] and others had noted that the curvature of wind, temperature, and specific humidity profiles are approximately equal to each other. This similarity is of both theoretical and practical importance; for example, it permits some inferences to be made about the heat flux (see Chapter 9). Lettau introduces two *Deacon numbers*, De and DE,

$$(7.14) \qquad \text{De} = -\frac{\partial \ln \dfrac{\partial \bar{u}}{\partial z}}{\partial \ln z} = -z \frac{\partial^2 \bar{u}}{\partial z^2} \bigg/ \frac{\partial \bar{u}}{\partial z}$$

$$(7.15) \qquad \text{DE} = -\frac{\partial \ln \dfrac{\partial T}{\partial z}}{\partial \ln z} = -z \frac{\partial^2 T}{\partial z^2} \bigg/ \frac{\partial T}{\partial z}$$

Formally, the similarity hypothesis is that De = DE.

Lettau develops a model for the lower atmosphere up to the level of a "constant" geostrophic wind. Although the details are largely outside the scope of what is usually included in micrometeorology, two of the predictions are of considerable interest:

1. An adiabatic atmosphere does not necessarily imply a logarithmic profile. When the geostrophic wind is low or when the surface is rough, the layer of constant shearing stress may be less than 10 meters in depth.
2. Large changes of De or DE with height (particularly a reversal from decreasing to increasing Deacon number) may be indicative of flux divergence of momentum and of heat, respectively.

It should be emphasized that the model has not yet been independently verified, although there are some suggestive results from the South Pole (see Chapter 15).

Monin and Obukhov [59] and Lettau [68] have suggested that equation (7.9) be generalized by adding an unspecified function ϕ:

$$(7.16) \qquad \frac{\partial \bar{u}}{\partial z} = u_* \phi / kz$$

Under adiabatic conditions, $\phi = 1$. It also follows from equation (7.10) that

$$(7.17) \qquad K_m = k u_* z / \phi$$

Monin and Obukhov reasoned dimensionally that ϕ is a function of $z/\mathfrak{L}$ where $\mathfrak{L}$ is a characteristic length.

Discussion of suggested universal forms for $\phi(z/\mathfrak{L})$ will be given in Chapter 9. However, two limiting cases can be described here. In any lapse rate for sufficiently small z, the buoyancy forces are negligible and the wind profile approaches the logarithmic case. Thus,

(7.18)
$$\lim_{z \to z_0} \phi(z/\mathfrak{L}) = 1$$

Furthermore, for $|z/\mathfrak{L}| \ll 1$, $\phi(z/\mathfrak{L})$ can be expanded as a power series, keeping only the first two terms:

(7.19)
$$\phi(z/\mathfrak{L}) = 1 + \alpha z/\mathfrak{L}$$

where α is to be determined empirically. Substitution of equation (7.19) in (7.16) leads to:

(7.20)
$$k\bar{u}/u_* = \ln z/z_0 + \alpha(z - z_0)/\mathfrak{L}$$

The limits of applicability of equation (7.20) and the value of the constant α have been the subject of some controversy. Monin and Obukhov's original estimate of 0.6 for α was for a time believed to be too low. On the basis of recent observations at Kerang, Australia,

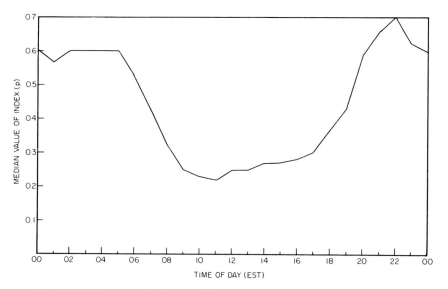

FIG. 23. Median values by hours of the day in summer of the power law exponent p at Douglas Point, Canada for the 20- to 80-ft layer [71].

however, Priestley [69] is in agreement with the value of 0.6. During superadiabatic rates, equation (7.20) is applicable only for $|z/\mathfrak{L}| < 0.03$ [70]; no limits have been established for inversion conditions.

Finally, it has been found empirically that wind profiles tend to obey a power law over a usefully thick layer:

$$(7.21) \qquad\qquad u/u_1 = (z/z_1)^p$$

where u_1 is the wind speed at some reference level z_1 and p is an exponent dependent on ground roughness and lapse rate. Median values of p obtained from the 20- and 80-ft winds at Douglas Point, Canada [71] are given in Fig. 23. As a general rule, the index is small during the day and large at night. An extensive summary of p-values for Brookhaven, Long Island has been given by DeMarrais [72].

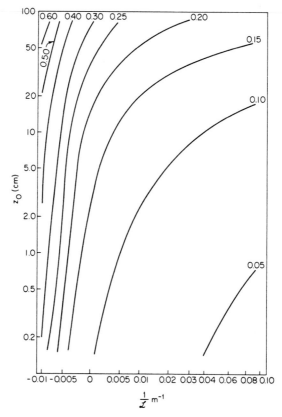

FIG. 24. Variation of the exponent p with roughness z_0 and the inverse of the Monin-Obukhov length $\mathcal{L}$ for the 11- to 46-meter layer; a zero value of $1/\mathcal{L}$ corresponds to adiabatic conditions [73].

Equation (7.21) is widely used by engineers. Panofsky *et al.* [73] have prepared a useful nomogram relating the index p to the roughness length z_0 and the inverse of the Monin-Obukhov length $\mathfrak{L}$ valid for the height range of 11 to 46 meters. The nomogram is reproduced in Fig. 24.

7.6. The Measurement of Mean Wind and Surface Shearing Stress

"Average" wind speeds are usually measured with cup anemometers. Each rotation of the cups produces either a count or voltage which is a measure of wind speed.

For careful profile measurements, the instruments should be calibrated at least once a day and should have relative errors of less than 5 cm/sec [74]. The disturbing effects of the sensor and support must also be recognized; if the anemometer is within 10 cm of the ground, the air flow may be seriously influenced by the instrument. At higher elevations, the effect of the mast and boom must be considered (see Chapter 11). In order to analyze curvature, an array of at least five sensors spaced at logarithmic height intervals is recommended.

The surface shearing stress may be measured with a floating lysimeter [26, 75, 76]. A sample of soil with representative vegetative cover is floated in oil, allowing the wind to act upon the surface. The frictional drag can be measured by using a spring mechanism. The spring deflection (or that of a balancing force that keeps the soil sample from moving) can be calibrated in terms of surface shearing stress. However, the accuracy of the measurements is not likely to be much better than $\pm 30\%$.

The quantities u_*, z_0, and d may be estimated from the wind profiles. In adiabatic conditions when the data fit equation (7.12), the parameters may be found graphically or by least square techniques. In other lapse rates, the method described by Dalrymple *et al.* [60] is recommended; it is based on Lettau's Deacon number theory. One of the practical experimental difficulties is that many sites are not in fact "infinite planes." A slight change in wind direction may have a large effect on the values of the roughness parameters.

8. Turbulence over Homogeneous Surfaces

8.1. The Nature of Turbulence

Turbulence is a very disorganized motion of the air, and it is not possible to predict the individual fluctuations from past history. Turbulence is therefore studied as a statistical random process.

In the wind tunnel, a mean wind $\bar{u}$ can be defined and measured at any point in the flow without too much difficulty. In the atmosphere, however, there have been many arguments concerning what constitutes mean flow and what constitutes turbulence. The controversy is now largely of historical interest. The present view is that a mean wind cannot be defined without specifying the time over which the average is obtained. The mean wind may be the average for a minute, an hour, a day, or a year; fluctuations about the mean are then called turbulence.

A knowledge of probability and advanced statistical theory is required in order to understand the present-day turbulence literature. However, much can be learned by watching a puff of smoke or by examining a wind record. Figure 25 gives three 1-min traces of elevation angle, obtained from a bidirectional vane positioned at 82 ft above the ground at Douglas Point, Canada. Figure 25 illustrates a property of turbulence called *intermittency*. The mean wind speed was 6.5 mph during the 7-min period and there was no detectable change in wind direction or lapse rate. Nevertheless, the character of the turbulence was quite different during the central minute. This is admittedly an extreme example but it does illustrate one of the difficulties in the statistical analysis of a wind record.

Although Fig. 25 demonstrates the turbulent nature of the wind, the record is of course smoothed by the limited response of the vane. There must be many additional fluctuations of high frequency that are not recorded.

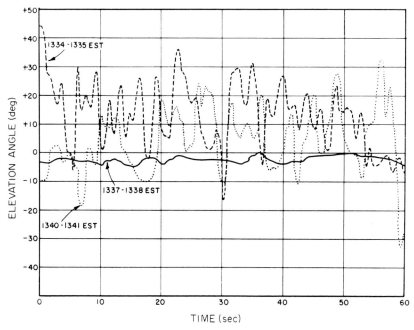

Fig. 25. Bidirectional vane record of elevation angle for three 1-min periods on July 19, 1962 at Douglas Point, Canada.

8.2. Some Definitions

The coordinate system used for specification of turbulence at a fixed point is as follows. The x-axis is chosen in the direction of the mean wind. The turbulent instantaneous values of the wind are $(\bar{u} + u')$, v', and w' in the x, y, z-directions, respectively. The mean wind is assumed to be blowing parallel to a horizontal surface, and the z-axis is in the vertical direction. By definition, $\overline{u'} = \overline{v'} = \overline{w'} = \bar{v} = \bar{w} = 0$.

As a first step in analysis, experimental data may be used to obtain the statistical *frequency distribution* of a variable. This is then compared with some relevant theoretical *probability density function*, such as the *normal* function (the bell-shaped Gaussian curve). It is important to distinguish between statistical and probability distributions, and this section considers the latter.

A widely studied parameter of a random process is the *variance*

$(\overline{u'^2})$, the average of the squares of the differences between individual values $(\bar{u} + u')$ and the mean $(\bar{u})$. Other parameters include quantities like $\overline{u'^3}$ (an index of skewness of a probability distribution), $\overline{u'^4}$ (an index of kurtosis or flatness of a probability distribution), and $\overline{u'w'}$ (the *covariance* between two random variables).

Some useful definitions are as follows:

1. The *intensity* or *level* or turbulence is given by the ratios

$$(\overline{u'^2})^{1/2}/\bar{u}, \; (\overline{v'^2})^{1/2}/\bar{u}, \; (\overline{w'^2})^{1/2}/\bar{u}$$

2. *Stationary turbulence* occurs when the probability distribution of the fluctuations is independent of time, i.e., there are no trends.

3. *Homogeneous turbulence* occurs when the probability distribution is independent of position in the fluid. It is usually assumed that surface roughness elements are randomly distributed and that turbulence is homogeneous in any horizontal plane. Turbulence is not homogeneous in the vertical direction because of shear.

4. *Isotropy* is defined by the relation

$$\overline{u'^2} = \overline{v'^2} = \overline{w'^2}$$

Stewart [77] has emphasized that it is impossible to have isotropy in shear flow. As a result, many of the theoretical results pertaining to isotropic turbulence do not apply to the surface boundary layer.

5. *Local isotropy* occurs when high frequency turbulence is isotropic, although low frequency fluctuations may not be. A limiting frequency must be specified when discussing local isotropy.

6. *Frozen turbulence*: Fluctuations (such as u') may be considered as an aggregate of sinusoidal waves, each with a frequency (n cycles/sec), a period ($2\pi/n$ sec), a wavelength (λ cm) and a wave number (K rad cm^{-1}). The dimensions of wave number are frequently abbreviated to K cm^{-1} and it is important not to confuse cycles cm^{-1}, also abbreviated as cm^{-1}, with rad cm^{-1} (2π rad = 1 cycle) [78].

The frozen turbulence hypothesis asserts that the wavelength is the product of period and mean wind:

(8.1) $$\lambda = 2\pi\bar{u}/n$$

(8.2) $$K = 2\pi/\lambda = n/\bar{u}$$

It is not implied that all the eddies move along with the mean flow velocity and without distortions. Individual eddies are in a state of constant growth and decay but, for low levels of turbulence, equations (8.1) and (8.2) are useful approximations.

8.3. The Problems of Normality and Intermittency in Shear Zones

Theory is simplified if probability distributions are *normal*, the defining parameters then being only the mean and the variance. Normality has been found to be a good approximation in the atmosphere over a homogeneous surface (e.g., Cramer [79]), with one exception. During superadiabatic conditions, a few rapidly rising currents of air are surrounded by large areas of slowly subsiding motion. The vertical fluctuations w' are then not *normally* distributed, although $\bar{w}$ may be zero.

Normality is also a good approximation in the wind tunnel, but Frenkiel [80] believes that the slight discrepancies are highly significant, being indicative of energy transfer and ultimate viscous dissipation from low to high frequency fluctuations.

Intermittency is a difficult problem on all scales of motion. For example, the frequency distribution of u' measured at 1-sec intervals over a month would probably be a smooth curve. However, there is much more turbulence during the day than at night. Hence, the distribution does not constitute a random sample but contains some degree of internal order. Statistical and probability theory is based upon the behavior of a *random* variable, and is not yet capable of including the effects of intermittency in any practical way. Stationary turbulence rarely exists in the atmosphere, although it may be a useful approximation for an hour or so when the wind speed and lapse rate are relatively constant.

Even in a wind tunnel where uniformity in the flow is most likely to be achieved, there is a nonuniform distribution of the fine structure of the turbulence. There appear to be negligible high frequency fluctuations over most of the flow, with relatively small regions of intense microscale turbulence. As a result, the viscous dissipation must be written as a fluctuating quantity, $(\bar{\varepsilon} + \varepsilon')$. Recent theoretical studies of the problem have been given [81–83]. It is expected that this will be a very active field of research in the next decade.

8.4. The Spectrum of Turbulence

The numerical value of a quantity such as $\overline{u'^2}$ is a measure of turbulent energy, and it is important to know how the energy is distributed among the various turbulent frequencies. A few large fluctuations might yield the same value of $\overline{u'^2}$ as would many small fluctuations. A spectrum analysis is therefore performed, which determines the contribution of various frequencies to the total energy.

The analogy with the spectral distributions of radiation (Chapters 2 and 3) is to be noted. The main difference is that electromagnetic waves preserve their wavelengths for long distances whereas turbulent fluctuations are chaotic and there is a continual transfer of energy through the spectrum. A further purely conventional distinction is that wave number $K(\text{cm}^{-1})$ or frequency n (cycles/sec) is used for the x-axis instead of wavelength λ. The y-axis is the spectral density $S(n)$, and the resulting spectrum is a curve with total area equal to $\overline{u'^2}$. The function $S(n)$ refers to any one-dimensional component of turbulence.

Several other coordinate axes are frequently used for displaying spectra:

1. The spectrum is sometimes "normalized" so that the area under the curve is equal to unity. This normalized function will be called $F(K)$ or $F(n)$. Thus,

$$(8.3) \qquad \int_0^\infty F(K)\, dK = 1 = \int_0^\infty S(K)\, dK/\overline{u'^2}$$

2. Since $\int F(K)\, dK = \int KF(K)\, d(\ln K)$, a plot of $KF(K)$ against $\ln K$ will preserve equality of areas in corresponding wave bands while at the same time compressing the scale; this is often an advantage because of the wide range of frequencies in the atmosphere.

3. Hinze [84] has shown that for local isotropy

$$(8.4) \qquad \varepsilon = 15\nu \int_0^\infty K^2 F(K)\, dK$$

Hence, a plot of $K^2 F(K)$ against K, or of $K^3 F(K)$ against $\ln K$, will show the contribution of different wave numbers to the energy dissipation rate, provided local isotropy exists.

4. A dimensionless frequency, $f = nz/\bar{u}$, is sometimes used; the spectra are then approximately independent of height.

Some typical spectra are given in Fig. 26 [85] obtained from a 10-min record of strong winds at Cardington, England.

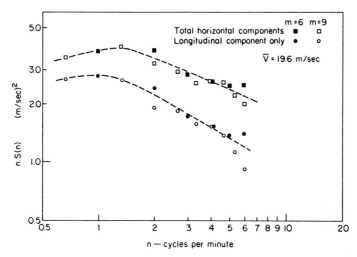

FIG. 26. An example of turbulence spectra; the horizontal gustiness at 15.3 meters at Cardington, England [85].

There are various methods of obtaining spectra from experimental data. A large computer or a commercially available wave analyzer may be used. Space limitations prevent inclusion of details here, but an excellent recent review has been given by Pasquill [86].

8.5. The Kolmogorov Similarity Theory

Some information about the shape of the spectrum is given by the Kolmogorov hypotheses [87]:

1. Energy enters the spectrum at relatively low frequencies and is transferred to higher and higher frequencies until it is finally dissipated. This is called a *cascade*.
2. All turbulent motions possess local isotropy in the high frequency end of the spectrum.
3. In the local isotropy region, the turbulent properties of the fluid are determined by v and ε. Dimensional arguments lead to

(8.5) $$F(K) = \varepsilon^{2/3} K^{-5/3} G(K, v)$$

where $G(K, v)$ is an undetermined dimensionless function.

4. At the low frequency end of the local isotropic range, viscosity has little effect on $F(K)$. This region is called the *inertial subrange* and in it, $G(K, \nu)$ is constant. Hence,

(8.6) $$F(K) = C\varepsilon^{2/3}K^{-5/3}$$

where C must be a universal constant. There are very few experimental determinations of C despite its obvious importance; for example, if C is known, equation (8.6) may be used to estimate ε from suitable experimental data. Grant *et al.* [88] have obtained a value for C of 0.47 ± 0.02, in the case of water turbulence in a Pacific coast tidal channel; the viscous dissipation rates ranged from 0.0015 to 1.02 cm^2/sec^3 (see discussion by Panofsky and Pasquill [78]).

The Kolmogorov theory is illustrated schematically in Fig. 27.

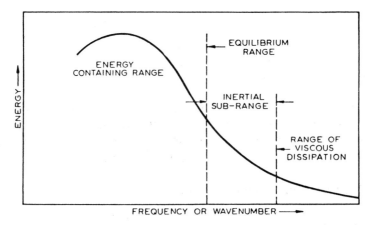

Fig. 27. Schematic representation of the spectrum of turbulence [86].

The model provides no information about the position or width of the wave number band for which equation (8.6) applies. Zubkovskii [89] has measured spectra at a height of 4 meters over short grass. His data suggest the following as a low frequency limit for local isotropy in the atmosphere:

(8.7) $nz/\bar{u} \simeq 0.08$ (unstable), $\quad nz/\bar{u} \simeq 0.2$ (neutral),

$$nz/\bar{u} \simeq 1.0 \text{ (stable)}$$

Gifford [90] points out that a $-5/3$ power law form for K is a necessary but not sufficient condition for an inertial subrange. Furthermore, a wavelength or frequency in the spectrum is not necessarily associated with a specific eddy size. The difficulty is that a three-dimensional spectrum cannot be measured experimentally; observed data are projections on a single axis, and this can be misleading.

A way of eliminating the effect of low frequencies is to position two wind sensors close together. The difference in wind speed fluctuations, $u'(x_2) - u'(x_1)$, where $x_2 - x_1$ is small, is a measure of the fine scale turbulence. This explains the use of the word "local" in the expression "local isotropy." A *structure function* S is defined as

$$(8.8) \qquad S = \overline{[u'(x_2) - u'(x_1)]^2}$$

and in the inertial subrange it may be shown dimensionally that

$$(8.9) \qquad S = C_1[\varepsilon(x_2 - x_1)]^{2/3}$$

The value of the constant has been estimated to be 1.6 by Obukhov and Yaglom [91] and 1.3 by Takeuchi [92].

Equations (7.7), (8.4), (8.6), and (8.9) permit the estimation of ε. Values have been given, for example, in [92–94]. As a general empirical rule, ε decreases with increasing height; it also varies as the third power of the wind speed at any height z; these relations might be anticipated from equation (7.7). Figure 28 [94] illustrates the behavior of ε.

8.6. The Effect of Sampling and Smoothing Times

Because a wind sensor does not respond completely to high frequency fluctuations, experimental records always include unavoidable smoothing. If the smoothing time is s, then it may be shown (see Pasquill [86], for example) that

$$(8.10) \qquad \overline{u_s'^2} = \overline{u'^2} \int_0^\infty F(n) \frac{\sin^2 \pi n s}{(\pi n s)^2}\, dn$$

For increasing s, $\overline{u_s'^2}$ approaches zero as more and more of the fluctuations are smoothed away. For decreasing s, on the other hand, $\overline{u_s'^2}$ approaches $\overline{u'^2}$.

Because a wind record is not of infinite length, some of the low frequency eddy energy will also be lost. If the length of record is

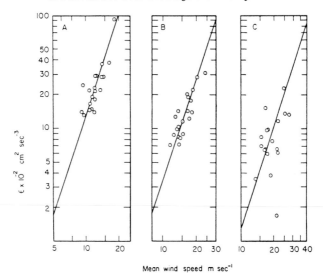

FIG. 28. Some estimates of the rate of viscous dissipation as a function of mean wind speed at Sale, Victoria, Australia at heights of 12, 64, and 153 meters from left to right [94].

T minutes, then it may be shown that

$$(8.11) \qquad \overline{u_T'^2} = \overline{u'^2} \int_0^\infty F(n)\left[1 - \frac{\sin^2 \pi nT}{(\pi nT)^2}\right] dn$$

The joint effect of a smoothing time s and a discrete length of record T is given by

$$(8.12) \qquad \overline{u_{T,s}'^2} = \overline{u'^2} \int_0^\infty F(n)\left[\frac{\sin^2 \pi ns}{(\pi ns)^2} - \frac{\sin^2 \pi nT}{(\pi nT)^2}\right] dn$$

These considerations are of major importance in the experimental determination of a spectrum. There are also applications in the estimation of diffusion of pollution from a stack (see Chapter 13).

8.7. Correlation Coefficients and the Scale of Turbulence

By definition, the autocorrelation coefficients are

$$(8.13) \qquad R(x) = \frac{\overline{u'(x_0)u'(x_0 + x)}}{\overline{u'^2}}$$

$$(8.14) \qquad R(t) = \frac{\overline{u'(t_0)u'(t_0 + t)}}{\overline{u'^2}}$$

It has been stated by Taylor [95] and verified by Panofsky *et al.* [96] that when the level of turbulence is low

$$(8.15) \qquad x = \bar{u}t \quad \text{and} \quad R(t) = R(x)$$

This consequence of the *frozen turbulence* hypothesis is illustrated in Fig. 29 [96].

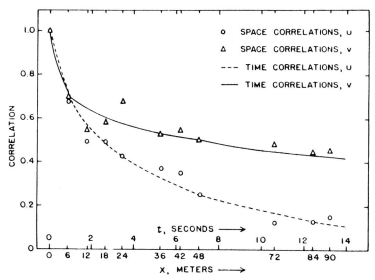

Fɪɢ. 29. Comparison between Eulerian space and time correlation functions at 2 meters at O'Neill, Nebraska under the assumption that $x = \bar{u}t$ [96].

Some general properties of $R(x)$ (space correlation) and $R(t)$ (time correlation) are as follows:

1. $R(0) = 1$, which follows directly from equation (8.13).
2. $\operatorname*{Lim}_{x \to \infty} R(x) \to 0$. $\operatorname*{Lim}_{t \to \infty} R(t) \to 0$. The physical justification for this follows from equation (8.20).
3. A correlation is an even mathematical function. Therefore, $R(-x) = R(x)$ and $R(-t) = R(t)$.
4. From points 1 and 3 it is evident that the correlation functions must have zero slope at the origin. Also, their second derivatives must be negative for small positive t or x.

5. Equation (8.9) may be used to deduce that in the inertial subrange

$$1 - R(x) \propto (\varepsilon x)^{2/3}$$

6. Taylor [95] has shown that the correlation function is the Fourier transform of the spectral function:

(8.16)
$$R(t) = \int_0^\infty F(n) \cos 2\pi nt \, dn$$

(8.17)
$$F(n) = 4 \int_0^\infty R(t) \cos 2\pi nt \, dt$$

7. Negative values are permissible for $R(t)$ and $R(x)$ but not for $F(n)$.

Another useful parameter in turbulence theory is the *scale of turbulence*, defined as follows:

(8.18)
$$\int_0^\infty R(x) \, dx = l_1 \qquad \text{(a length)}$$

(8.19)
$$\int_0^\infty R(t) \, dt = l_2 \qquad \text{(a time)}$$

Substitution of $n = 0$ in equation (8.17) leads to

(8.20)
$$F(0) = 4 \int_0^\infty R(t) \, dt = 4l_2$$

Thus the scale of turbulence is equal to one-quarter of the zero intercept of the spectral function. Since a zero frequency corresponds to a harmonic of infinite wavelength, i.e., a straight line, it follows that the scale of turbulence must be finite; otherwise the eddy energy in the atmosphere would be infinite.

8.8. Cross-Spectrum Analysis

Quantities such as $\overline{u'w'}$ are of importance in a fluid. Since the wind near the ground increases with height, small values of u' at any given point are generally associated with upward moving air while large values of u' indicate subsidence. A nonzero value of $\overline{u'w'}$ is therefore probable. The question arises as to which frequencies in the u' and w' spectra contribute to the magnitude of $\overline{u'w'}$. What, for example, is its value in the inertial subrange?

The standard way of analysis of correlations between two meteorological series is cross-spectrum analysis [97] yielding *cospectrum*, *quadrature spectrum*, and *coherence*. The cospectrum $Co(n)$ is a display of $Co(n)$ versus frequency n, where

$$(8.21) \qquad \int_0^\infty Co(n) \, dn = \overline{u'w'}$$

There is always a possibility that two time series will appear to be uncorrelated because the correlation coefficient changes sign through the spectra. In order to examine this possibility, the quadrature spectrum $Q(n)$ is calculated. The spectrum of one variable is obtained, and each frequency is lagged by one-quarter period before commencing cross correlation with the other variable.

The coherence $CH(n)$, is defined as follows:

$$(8.22) \qquad CH(n) = \frac{Q^2(n) + Co^2(n)}{F_{u'}(n)F_{w'}(n)}$$

The coherence varies from 0 to 1 and is an index of the exactness of the relation between the two variables for various frequencies.

8.9. Shearing Stress in Terms of Eddy Fluctuations

It can be shown by a very general argument, that when density fluctuations are neglected,

$$(8.23) \qquad \tau/\rho = u_*^2 = -\overline{u'w'}$$

The proof is as follows. Suppose that $E = \bar{E} + E'$ is the amount per unit mass of fluid of any transferable conservative quantity such as water vapor, heat, or momentum. Assume that $\bar{v} = \bar{w} = 0$, that $\bar{E}$ is homogeneous in the horizontal, and that molecular motions may be neglected in comparison with turbulent transfer. The instantaneous flux of $E = F_E = -\rho E w'$;

hence, the mean flux of $E = \bar{F}_E = -\overline{\rho E w'}$

$$= -\overline{(\bar{\rho} + \rho')(\bar{E} + E')w'}$$

Hence, since $\overline{w'} = 0$,

$$(8.24) \qquad \bar{F}_E = -\bar{\rho}\,\overline{E'w'} - \bar{E}\,\overline{\rho'w'} - \overline{\rho'E'w'}$$

Equation (8.23) follows when the property being transferred is identified as momentum, i.e., $E = u$, and when density fluctuations are neglected, i.e., the last two terms in equation (8.24) are assumed to be small. This latter assumption is not immediately obvious although widely accepted. It will apply in an incompressible fluid but has not been adequately demonstrated for the atmosphere. However, the difficulty is overcome by using a hot-wire anemometer [98], which measures $\rho w'$ directly, in conjunction with a cup or propeller type for u'.

Equation (8.23) provides another method of obtaining experimental "estimates" of shearing stress, and of related parameters such as momentum diffusivity and viscous dissipation. The various methods that may be used to estimate shearing stress have been discussed recently [99–101].

8.10. The Lagrangian Reference Frame

Most meteorological measurements are obtained from fixed points in space, in what is called the *Eulerian* reference system. The alternative is to follow the behavior of a single weightless particle or fluid element as it moves through the atmosphere. This is known as the *Lagrangian* reference frame.

There are many difficulties in obtaining Lagrangian measurements. Soap bubbles, zero-lift balloons, and tetroons have been used but there are relatively few experimental data. Nevertheless, many of the fundamental theorems of fluid mechanics are cast in the Lagrangian system, and much attention has been directed recently toward relating Lagrangian and Eulerian reference frames.

For stationary turbulence, Lagrangian parameters are independent of the initial time at which the fluid element is followed. Furthermore, quantities such as $\bar{u}$, $\overline{u'^2}$, $\overline{v'^2}$, $\overline{w'^2}$, and $\overline{u'w'}$ are the same in the Eulerian and the Lagrangian systems, although spectra and correlation functions are different.

Hay and Pasquill [102] have made the simple assumption that

$$(8.25) \qquad R_L(\xi) = R_E(t) \qquad \text{when } \xi = \beta t$$

where $R_L(\xi)$, $R_E(t)$ are the Lagrangian and Eulerian correlation functions and β is the ratio of the Lagrangian to the Eulerian time scales. It follows from equation (8.17) that

$$(8.26) \qquad F_L(n) = \beta F_E(n)$$

The relations are given schematically in Fig. 30 [86]. Experimental data support equations (8.25) and (8.26) as useful assumptions, but there is still some question regarding the value of β, and whether it is constant with stability or turbulence intensity. Hay and Pasquill

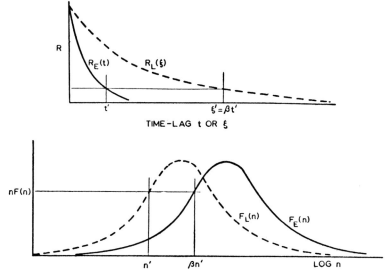

FIG. 30. Schematic model for scale relations between Lagrangian and Eulerian correlation functions and spectra [86].

suggest a value for β of 4; Angell [103] obtains values of 1 and 7 for turbulence intensities of 0.35 and 0.05, respectively, based on 25 daytime experiments. Wandel and Kofoed-Hansen [104] have considered the special case of homogeneous, isotropic turbulence and infer that

$$(8.27) \qquad \beta \simeq \sqrt{\pi}/4i$$

8.11. The Measurement of Turbulence

A number of excellent textbooks on the theory of turbulence are available, including those by Hinze [84] and Lumley and Panofsky [105]. It has been possible to consider only a few selected topics in this chapter, but one of the most important aspects, experimental measurements, should not be overlooked.

A bidirectional vane is commonly used to measure the elevation and azimuth angles of the wind. Very careful attention to design is required in order to overcome the problem of resonance between the turbulence and the natural period of vibration of the vane. Wind direction fluctuations measured with a standard anemometer will "overshoot" by more than 100% at certain resonant frequencies.

The response of a vane is specified by a *distance constant*, the product of the time constant of the vane and the wind speed. The distance constant is approximately constant for any given vane [106, 107].

The three components of turbulence are measured in the wind tunnel with a hot-wire anemometer. The method is not very successful in the atmosphere because of calibration drift caused by the deposition of impurities on the sensor. Much interest is therefore being shown in the *sonic* anemometer, which operates on the principle that the velocity of a sound wave in a moving medium is equal to the velocity of sound with respect to the medium plus the velocity of the medium [26].

Note Added in Proof

The experimental value of 0.47 for the constant C in equation (8.6) (see page 72) has recently been verified theoretically by R. H. Kraichnan [(1965) *Phys. Fluids* **8**, 995–997].

9. Turbulent Transfer of Heat from Homogeneous Surfaces

9.1. The Assumption of Constant Vertical Heat Flux

A fundamental assumption in micrometeorology has been that the turbulent vertical heat flux Q_H is constant with height in the surface boundary layer. The justification was that air temperatures remain relatively constant with time for an hour or so in the early afternoon, and for a few hours at night. Vertical flux divergence of heat must be negligible during those periods.

Air temperatures are controlled by both radiative and turbulent processes, and Funk [18] has shown that radiative flux divergence is important at night. Constant temperatures imply merely that radiative and turbulent flux divergences are in equilibrium; no indication is given concerning the magnitude of the terms. If a similar situation exists during the day, the assumption of constant Q_H is in serious doubt. This is particularly true when haze or water vapor concentrations are relatively high.

A large body of internally consistent theory has been developed using the hypothesis of *constant* Q_H. These physical models will be described first; the relevance to an atmosphere in which radiative transfer is important can then be considered.

9.2. The Monin-Obukhov Length and the Richardson Number

Monin and Obukhov [59] used a dimensional argument to deduce the nature of the length $\mathfrak{L}$ of the function $\phi(z/\mathfrak{L})$ of equation (7.16). When buoyancy forces exist, the character of the flow depends no longer upon the Reynolds number but upon u_*, Q_H, ρ, c_p, g, and T. These may be combined dimensionally to yield a length $\mathfrak{L}$:

(9.1)
$$\mathfrak{L} = -u_*^3 T \rho c_p / k g Q_H$$

The negative sign and von Kármán's constant have been added for later convenience. The quantity $\mathfrak{L}$ is infinite when the heat flux is zero. It is negative when the lapse rate is superadiabatic and positive during inversions.

Alternatively, the *Richardson flux* number, Rf, may be used. It is related to $\mathfrak{L}$ by the equation

$$(9.2) \qquad\qquad Rf = z/\mathfrak{L}\phi$$

When the lapse rate approaches the adiabatic, the function ϕ approaches unity as was indicated in equation (7.18). Then

$$(9.3) \qquad\qquad Rf \simeq z/\mathfrak{L}$$

The flow is preferably characterized by $\mathfrak{L}$ rather than by Rf because the latter varies with height. The similarity theory is that the surface boundary layer expands or contracts according to the size of $\mathfrak{L}$.

A thermal diffusivity K_H is next introduced. By analogy with soil heat diffusivity K_G and equation (5.2), but using the potential temperature of equation (6.1),

$$(9.4) \qquad\qquad Q_H = -\rho c_p K_H (\partial\theta/\partial z)$$

It is self-evident from equation (9.4) that the vertical heat flux is zero when the potential temperature gradient is zero (but see Section 9.4).

Finally, the *Richardson number*, Ri, is defined:

$$(9.5) \qquad\qquad Rf = (K_H/K_m)Ri$$

Substitution of equations (7.6) and (9.4) in (9.5) yields the more usual form for the Richardson number:

$$(9.6) \qquad\qquad Ri = \frac{g(\partial\theta/\partial z)}{T(\partial\bar{u}/\partial z)^2}$$

The Richardson number has long been used as an index of stability. Richardson [108] showed that the flux form, Rf, is the ratio of the rate at which buoyancy forces extract energy from turbulence to the rate at which it is supplied by wind shear.

Although Rf is a fundamental dimensionless number, it cannot be calculated readily from experimental data. The quantity, Ri, is therefore more commonly used while in some cases a simplified parameter, the *stability ratio SR*, is quite satisfactory.

$$(9.7) \qquad\qquad SR = \frac{Tz_2 - Tz_1}{\bar{u}^2}\, 10^5 \quad (°C \text{ sec}^2/cm^2)$$

where $\bar{u}$ is measured at a height equidistant from z_1 and z_2 on a logarithmic scale. Wind speed can be measured with higher relative percentage accuracy than wind shear. Since the two vary with lapse rate in approximately the same fashion, data are often usefully stratified according to *SR*.

9.3. The Ratio of Diffusivities K_H/K_m

It was at one time believed that the ratio K_H/K_m was constant and probably equal to unity. However, Ellison [109] pointed out that it is quite possible for a displaced bubble of air to return to its original position without mixing with its environment. In these circumstances, momentum will be exchanged through the mechanism of pressure fluctuations whereas there will be little transfer of heat. Ellison predicted that during very stable conditions, both K_H and the ratio K_H/K_m would become vanishingly small. However, when vigorous mixing is taking place, e.g., during strong forced convection, the pressure fluctuation mechanism is hardly likely to be of importance (except possibly at heights of order-of-magnitude z_0; see Section 16.7).

There is general agreement (e.g., Robinson [110]) that when the lapse rate is superadiabatic, heat is transferred by fluctuations of lower frequency than is momentum. The effect on the ratio K_H/K_m is not clear. Experimental data (e.g., Swinbank [98]) suggest that the ratio is greater than unity. However, inferences from the measurements are all contaminated by the possible intrusion of radiative flux divergence. A number of results have been summarized by Senderikhina [111] in Table VI using the inverse ratio K_m/K_H.

The ratio K_m/K_H had a median value of 0.68. The 78 observations

TABLE VI. Values of the ratio K_m/K_H [111].

| Source | No. of observations | Values of ratio K_m/K_H | | |
		Median	Upper quartile	Lower quartile
Rider	27	0.72	0.76	0.58
Perepelkina	20	0.86	1.48	0.28
Gurvich and Zwang	31	0.56	0.81	0.37
All cases	78	0.68	0.88	0.46

included both positive and negative lapse rate cases but the scatter was too great to determine an empirical relation between the ratio and lapse rate or Richardson number.

Swinbank [112] has analyzed recent data from an "infinite plane" at Kerang, Australia. He concludes that in superadiabatic conditions: (a) the ratio K_H/K_m increases with height; (b) the ratio also increases with increasing $-\mathfrak{L}$; for the 0.5 to 1.0-meter layer, the value changed from 1.1 at $\mathfrak{L} \simeq -40$ to 2.5 at $\mathfrak{L} \simeq -4$.

The present position is far from satisfactory, and new experiments are required in dry, haze-free environments. The importance of the problem stems from the fact that equations (7.6) and (9.4) are sometimes combined to give

$$(9.8) \qquad Q_H = -\rho c_p u_*^2 (K_H/K_m)\left(\frac{\partial \theta}{\partial z}\right)\bigg/\frac{\partial \bar{u}}{\partial z}$$

With the additional dubious assumption that the diffusivities are equal, equation (9.8) is then used to compute Q_H. This method is to be discouraged.

9.4. Daytime Turbulent Heat Fluxes

The problem of predicting the onset of free convection has not been solved theoretically. It does not occur at the point where the lapse rate becomes positive because viscous and shearing forces tend to damp away convection cells. Priestley [93] has found experimentally that the transition occurs at a Richardson number of about -0.03.

In the lowest few meters on a sunny afternoon, three distinct regimes can be predicted [113]:

1. Mechanical turbulence is dominant very near the ground ($-\text{Ri} < 0.03$). When equations (7.16) and (7.19) are substituted in (9.8), it follows that:

$$Q_H = -\rho c_p u_*(K_H/K_m)(\partial \theta/\partial z)kz/(1 + \alpha z/\mathfrak{L})$$

With the additional assumption that (K_H/K_m) is constant, it follows that the temperature gradient varies inversely as the height in the region of forced convection.

2. There is an intermediate layer of free convection. Priestley [114] used a dimensional argument to derive an expression for Q_H during conditions of strong instability ($-\text{Ri} > 0.03$, experimentally) when the heat flux becomes independent of wind shear.

Assuming that Q_H depends upon ρ, c_p, g/T, z, and $(\partial\theta/\partial z)$, it follows that

(9.9)
$$Q_H = h\rho c_p (g/T)^{1/2} z^2 \left|\frac{\partial\theta}{\partial z}\right|^{3/2}$$

where h is a constant to be determined experimentally. The value of h was estimated to be about 0.9 [114], but recent measurements at Kerang, Australia [69] have yielded a value of 1.27. It follows from equation (9.9) that

(9.10)
$$\frac{\partial\theta}{\partial z} \propto z^{-4/3}$$

3. There is an upper layer of natural convection commencing at a height roughly equivalent to the Monin-Obukhov length $\mathcal{L}$. Townsend [113] has shown how the laboratory expression for natural convection in the absence of wind may be modified for the atmosphere.

(9.11)
$$-Q_H = g^{1/2} z^3 \left(\frac{1}{T}\right)^{3/2} \left(\frac{\partial\theta}{\partial z}\right)\bigg/0.44\mathcal{L}$$

Hence,

(9.12)
$$\frac{\partial\theta}{\partial z} \propto z^{-2}$$

Townsend suggests that the intermediate region of free (or mixed) convection arises in the atmosphere but not in the laboratory because of the presence of wind shear, which indirectly affects the rising buoyant plumes through turbulent interactions, even though the wind shear does not appear directly in equation (9.9).

These results provide a method for estimating Q_H during daylight hours. The first step is to examine vertical temperature profiles and to determine the layers where temperature gradients vary as z^{-1}, $z^{-4/3}$, and z^{-2}. The appropriate equations for Q_H can then be used. Unfortunately, there are not many experimental sites where the surface is sufficiently homogeneous for the demarcation among the three power laws to be determined.

When an inversion caps a superadiabatic surface layer, there will be *penetrative convection* of the stable layer by buoyant plumes or bubbles. The mean flow of heat may then continue to be upward (against the potential temperature gradient), in contrast with ideas advanced in Section 1.2.

9.5. Nighttime Turbulent Heat Fluxes

When the air becomes very stable, turbulence decreases and ultimately the flow becomes laminar. The *critical Richardson flux number*, Rf_{crit}, is defined as that value of Rf associated with the transition from turbulent to laminar flow. It is believed to have a numerical value between 0.14 and 0.5 [109, 115]. Stewart discusses the implications with respect to momentum and heat transfer. He concludes that no present theory is capable of predicting fluxes in very stable situations. Vertical fluxes are probably due to internal gravity waves, a subject about which "disappointingly little" is known.

Fortunately, downward heat fluxes at night are small in comparison with daytime upward heat exchanges. The present state of the theory is such that no quantitative method can be recommended for estimating Q_H during strong inversions.

9.6. Viscous Dissipation and the Diabatic Wind Profile

In the absence of buoyancy forces, it was possible to find dimensional expressions for the rate of energy production ($u_*^2 \partial \bar{u}/\partial z$) and for the rate of viscous dissipation (u_*^3/kz). For the diabatic case, the rate of energy production must include an additional buoyancy term. A dimensional argument [62] leads to a total production of

$$u_*^2 \frac{\partial \bar{u}}{\partial z} - Au_*^3/k\mathfrak{L}$$

where A is a measure of the relative efficiency of the buoyant and shear forces in producing turbulent energy. Although there has been much speculation, it is not yet known whether the quantity A is a constant or is a function of $z/\mathfrak{L}$.

Another fundamental difficulty is that the dimensional form for the rate of viscous dissipation has not been settled. Consequently, it is not possible to present a universal form for the diabatic wind profile. Some information, however, is available from three special limiting cases [116]:

1. In an adiabatic atmosphere, $\phi(z/\mathfrak{L}) = 1$, and the wind profile is logarithmic, equation (7.11).

2. In strong stability, ϕ/z can no longer be a function of the height z. Hence, the wind must increase linearly with height.
3. In free convection, the form of the temperature gradient was given by equation (9.9). Taylor [70] postulated similarity of wind and temperature profiles, and verified his assumption experimentally. This leads to

$$(9.13) \qquad \phi = c_1 |z|\mathfrak{L}|^{-1/3} \qquad \frac{\partial \bar{u}}{\partial z} = c_2 |z|\mathfrak{L}|^{-4/3}$$

An empirical relation that is at least correct for these three limiting cases has been proposed independently by Kazansky and Monin [117], Ellison [109], Yamamoto [118], Panofsky [119], and Sellers [120]. It may be written in three equivalent forms and is sometimes called KEYPS:

$$(9.14) \qquad \phi^4 + A \frac{z}{\mathfrak{L}} \phi^3 = 1$$

$$(9.15) \qquad K_m = \left[u_*^2 \frac{\partial \bar{u}}{\partial z} - A u_*^3 / k\mathfrak{L} \right]^{1/3} (kz)^{4/3}$$

$$(9.16) \qquad K_m = k u_* z [1 - \mathrm{Rf}/\mathrm{Rf}_{\mathrm{crit}}]^{1/4}$$

Equations (9.14) to (9.16) have been employed with considerable mathematical versatility to produce a number of "universal" wind profiles, e.g., Yamamoto [118]. The essential difficulty is that the quantity A may be a function of stability, i.e., of $z/\mathfrak{L}$. No final judgement can be passed until more is known concerning the relative efficiency with which buoyant and shear forces produce turbulent energy.

Swinbank [112] has deduced from some intuitive assumptions that

$$(9.17) \qquad \frac{\partial \bar{u}}{\partial z} = \frac{u_*}{k\mathfrak{L}} [1 - \exp(-z/\mathfrak{L})]^{-1}$$

Expanding the right side of equation (9.17) as a series in $z/\mathfrak{L}$ leads to equation (7.20) with a value of α of 0.5. The second coefficient is 0.083 while succeeding ones are very small. Experimental data from Kerang, Australia, which included independent estimates of Q_H, show good agreement with values of Q_H obtained from equation (9.17). Nevertheless, Lettau [66] argues that the model requires a fuller examination of the second derivatives of wind speed, e.g., through the Deacon number of equation (7.14).

9.7. The Eddy Correlation Method for Measuring Heat Flux

From equation (8.24) it may be shown that

$$(9.18) \qquad Q_H = c_p \overline{\rho w' T'} \simeq c_p \rho \overline{w' T'}$$

The principal difficulty in exploiting equation (9.17) is in matching the response of wind and temperature sensors. However, the problem has been largely overcome by recent instrumental developments in Australia and the U.S.S.R. [121–123]. Dyer's instrument, *the evapotron*, measures vertical fluxes of both heat and water vapor. He has shown [124] that if the time constant of the sensors is T, the fluxes are almost completely recovered provided that the numerical value of $T\bar{u}/z$ is less than 0.15. When the parameter increases to 0.5, the measured fluxes are only about 90% of those obtained from the energy balance equation because the contributions to the correlations by the high frequency fluctuations are lost.

It is of some importance to know what frequencies are responsible for vertical fluxes of various properties. Is heat being transferred by the same size eddy as is momentum? Because of instrumental differences and because it is difficult to find a homogeneous "infinite plane," no clear picture has yet emerged from the experimental work. However, a few general statements can be made:

1. The energy in turbulence spectra and cospectra shifts to lower frequencies with increasing distance from the ground and/or greater instability.
2. In the daytime, heat is transferred at lower frequencies than is momentum.
3. There has been a widespread belief in the existence of a *laminar sublayer*, a thin film of air near the ground where heat and momentum are transferred entirely by molecular action. The reasoning has been that eddies with vertical dimensions greater than the height above the ground cannot exist. However, Sternberg [125] suggests that the wind flow is turbulent right down to the interface. Although Sternberg retains the laminar sublayer, he believes that large eddies occasionally penetrate the layer from above. This mechanism may be responsible for vertical fluxes but its exact effect remains to be elucidated. Sternberg also suggests that the large eddies move downstream with a speed characteristic of the upper regions of the shear zone; at lower levels, therefore, they

move more quickly than the local mean wind would appear to indicate, in contradiction to the *frozen turbulence* hypothesis, which consequently must be restricted to wavelengths smaller than the height above ground. Since heat transfer begins at the interface, physical processes taking place in the lowest few millimeters or centimeters are of importance, although this is a region where it is almost impossible to obtain experimental data. Controlled wind tunnel studies are therefore to be encouraged.

4. It has been verified experimentally [126] that there is an inertial subrange of temperature fluctuations that conforms to a $-5/3$ power law similar to that given by equation (8.6). Taylor [127] has used the result in developing a general equation of heat balance, neglecting radiative transfer. An order-of-magnitude examination of the various terms in the equation leads to a simple and very useful approximation:

$$(9.19) \qquad \varepsilon_T = -\frac{Q_H}{\rho c_p} \frac{\partial \theta}{\partial z}$$

where ε_T is the rate of dissipation of turbulent heat fluctuations. Equation (9.18) may be used to determine Q_H from measurements of lapse rate and of high frequency temperature fluctuations.

9.8. The Effect of Radiative Flux Divergence on Heat Transfer

The observed temperature structure of the surface boundary layer is the integrated result of turbulent mixing and radiative transfer. One of the unsolved problems of micrometeorology is the determination of the relative importance of the two components. A practical application arises from the fact that it is frequently required to interchange the diffusivities of momentum, heat, and water vapor. If experimental estimates of K_H include the effect of radiation, then that diffusivity may be different from the other two, apart from other reasons discussed elsewhere.

During intense inversions it is generally agreed that temperature profiles are controlled mostly by radiation. This is believed to be the reason, for example, why the minimum temperature often occurs a few centimeters above the ground. After the inversion has developed, the ground begins to receive heat by long-wave radiation from the sky; in combination with Q_G and Q_H, the energy gain at the surface may become sufficiently large to reduce the long-wave upward radiative loss, with a resulting increase in ground temperature.

Gaevskaya *et al.* [128] approach the problem through a general form for the heat transfer equation:

$$(9.20) \qquad \frac{\partial T}{\partial t} = \frac{\partial}{\partial z}\left(K_H \frac{\partial T}{\partial z}\right) + \frac{1}{\rho c_p} \frac{\partial Q_n}{\partial z}$$

where K_H is restricted to the turbulent component of vertical heat flux. The similarity of equation (9.19) with equations (5.6) and (5.8) is to be noted.

An approximate solution is obtained based on a number of simplifying assumptions. It is found that temperature changes caused by radiation and by turbulence are not additive, i.e., a separate calculation of either component has no meaning. Some results are given in Table VII, assuming an initially isothermal atmosphere and soil, a

TABLE VII. Calculated cooling of an isothermal atmosphere in 4 hr ($^\circ$C) [128].

Height, cm	Radiative cooling	Turbulent cooling	Sum of radiative and turbulent cooling	Radiative and turbulent cooling by eq. (9.19)
0	−1.78	−4.29	−6.07	−3.76
2	−1.78	−3.11	−4.89	−2.78
4.8	−1.78	−2.82	−4.70	−2.54
24.0	−1.78	−2.26	−4.04	−2.06
48.0	−1.78	−2.05	−3.83	−1.89
200.0	−1.78	−1.52	−3.30	−1.45
480.0	−1.78	−1.22	−3.00	−1.20

constant radiative flux divergence, and using equation (5.6) for soil heat flux. Although the simplifying assumptions are not very realistic, the method is promising.

Kraus [16] approaches the problem from a somewhat different point of view, although he does begin with equation (9.20). Whereas Gaevskaya *et al.* assumed an empirical form for K_H and attempted to solve the nonstationary heat exchange problem, Kraus uses observed values of $\partial T/\partial t$ and radiation-chart estimates of $\partial Q_n/\partial z$ to deduce $\partial Q_H/\partial z$. Finally, he is able to obtain an estimate of the relative error in the austausch coefficient ρK_H that would be made by neglecting radiative flux divergence. Some results are given in Table VIII.

Elliott [129] has estimated the radiative and turbulent flux divergences at O'Neill, Nebraska [26]. He believes that daytime radiative

TABLE VIII. Estimates of flux divergence [16].

Place	Seabrook, N.J.	Seabrook, N. J.	Perlacher Forest
Time	1300–1400	1900–2000	1730
z_0, cm	10	10	50
z, cm	640	640	600
$\int_{z_0}^{z} \rho c_p \dfrac{\partial T}{\partial t} \, dz$ mcal/cm^2 min	2	-5	-15
$\int_{z_0}^{z} \dfrac{\partial Q_N}{\partial z} \, dz$ mcal/cm^2 min	14	-3	-7
Relative error, %, in austausch coefficient	2	7	50

flux divergence, although able to produce large temperature changes, has little effect on the constancy of Q_H in the lowest few meters (because Q_H is a relatively large quantity). Although this assessment is likely to be correct, the flux divergences are undoubtedly important in determinations of the constants in dimensional equations and in questions involving the ratios of diffusivities.

10. Evaporation from Homogeneous Surfaces

10.1. The Evaporation Process

Water molecules are continually moving across the earth-atmosphere interface. If the number of molecules escaping from the ground is greater than the number returning, water is evaporating. Most of the transfer takes place within 1 to 2 molecular free paths of the surface, and the violent activity in this region is not always realized. Hickman [130] estimates that 2 to 3 kg of water are moving across the interface in each direction every second over each square meter of an open water surface. The net exchange is of course an insignificant fraction of the total.

Continuous evaporation requires the removal of water vapor from just above the surface. Without turbulent diffusion, the air near the ground would become saturated and evaporation would stop.

A second requirement for evaporation is a heat source. The surface is cooled by the evaporating water, thus reducing the saturated vapor pressure. Without a heat source, equilibrium would soon be reached and evaporation would cease.

A third important factor is the physical or chemical properties of the underlying liquid (never pure water in nature). For example, Sechrist [131] has evidence suggesting that water evaporates more rapidly when it contains dissolved carbon dioxide.

Some attempts have been made by physical chemists to study evaporation on the molecular scale. The micrometeorologist, however, must be content with experimental measurements a few centimeters or meters above the ground.

10.2. Some Formal Relations

A vertical gradient of water vapor in the atmosphere implies a vertical flux. The flux divergence is usually assumed to be negligible; there is no complicating contribution from radiation as in the case of Q_H. It is then possible by analogy with equation (5.2) to write

$$(10.1) \qquad Q_E = LE = -L\rho K_E \frac{\partial q}{\partial z}$$

where L is the latent heat of evaporation and K_E is *water vapor diffusivity* by definition.

The usefulness of equation (10.1) is restricted since little is known about the behavior of K_E. The classical approach was to assume adiabatic conditions. Integration of equation (10.1) from z_1 to z_2 leads to

$$E = -\rho(q_2 - q_1) \bigg/ \int_{z_1}^{z_2} (K_E)^{-1} \, dz$$

If $K_E = K_m = ku_* z$,

$$E = -\rho(q_2 - q_1)ku_*/\ln(z_2|z_1)$$

The friction velocity may be eliminated by substitution of equation (7.11) (with the integration performed from z_1 to z_2 rather than z_0 to z).

$$(10.2) \qquad E = -\frac{k^2\rho(q_2 - q_1)(u_2 - u_1)}{(\ln z_2|z_1)^2}$$

This is the familiar Thornthwaite and Holzman equation [132].

If vertical temperature gradients are averaged over a relatively long period, such as a month, the mean lapse rate may be quite close to the adiabatic. However, the application of equation (10.2) *after* the data have been averaged is to be discouraged. Mean values do not in general produce mean rates of evaporation.

An extension to nonadiabatic conditions is possible by combining equations (7.6) and (10.1). It may be derived that

$$(10.3) \qquad E = -\rho u_*^2 (K_m/K_E) \frac{\partial q}{\partial z} \bigg/ \frac{\partial \bar{u}}{\partial z}$$

However, the determination of u_* is difficult and not yet feasible on a routine basis [99]. Furthermore, there is the same problem as in the case of equation (10.2); the value of the ratio K_m/K_E is not well established.

Another approach is possible through the Bowen ratio R. From equations (9.4) and (10.1):

(10.4)

$$R = Q_H/Q_E = c_p K_H \frac{\partial \theta}{\partial z} \bigg/ L K_E \frac{\partial q}{\partial z} \simeq c_p K_H (T_2 - T_1) \bigg/ L K_E (q_2 - q_1)$$

If the ratio K_H/K_E is known, then a good estimate of the Bowen ratio could be found from measurements of temperature and specific humidity at two levels. It would then be possible to obtain either Q_H or Q_E from equation (1.1).

10.3. The Ratio of Diffusivities

The relative magnitudes of K_m, K_H, and K_E is evidently a question of some importance. It should be emphasized at the outset, however, that there are no valid theoretical reasons for attaching any physical significance to the turbulent diffusivities. The question to be answered, instead, is whether the approach provides a useful empirical method for determining vertical fluxes.

Ellison [109] reasoned that the turbulent transfer processes for heat and vapor must be similar because both require actual mixing. Momentum, on the other hand, can be exchanged by pressure fluctuations without actual mixing. Unfortunately, thermal diffusivity includes a radiative component; hence, if Ellison is correct, $K_H > K_E$ when the quantities are obtained from observed profiles, the difference being dependent upon the magnitude of the radiative flux divergence.

Suitable data for testing the predictions from independent measurements of shearing stress and water vapor flux are almost nonexistent. Senderikhina [111] used 39 observations by Rider [67] to obtain a median value of 0.85 for the ratio K_m/K_E with upper and lower quartile values of 0.98 and 0.74, respectively. No separation of cases according to lapse rate was possible. Senderikhina concluded that the available data supported an inequality (except during strong inversions):

(10.5) $K_m/K_H < K_m/K_E < 1$

The inequality is not firmly established but it does agree with Ellison's views.

Other estimates of the diffusivity ratios have been given recently [133] based on data from Project Green Glow at Hanford, Washington and from O'Neill, Nebraska [26]. The values were not obtained from fundamental eddy correlations but were derived from an indirect use of the energy balance equation. Nevertheless, the results (at least for lapse conditions) are in qualitative agreement with Senderikhina (Table IX).

TABLE IX. Empirically determined values of the diffusivity ratios [133].

	Daytime		Nighttime	
	K_H/K_m	K_E/K_m	K_H/K_m	K_E/K_m
Green Glow	2.92	1.53	0.51	0.45
O'Neill	1.54	1.13	0.48	0.28

A. J. Dyer (1965, to be published) has recently analyzed daytime data from Hay, Australia which clearly establish that $K_H = K_E$ in a relatively dry and haze-free atmosphere. The Bowen ratio approach, equation (10.4), is therefore correct in lapse conditions for environments where radiative flux divergence may be neglected. However, more experiments are required during high humidity and haze, as well as when inversions are present.

10.4. Some Recent Experimental Data

An important study by Crawford [134] relates fluxes determined by the gradient method with independent measurements of evapotranspiration obtained from the large weighing lysimeter at Davis, California.

A nondimensional form $\overset{*}{E}$ of the evaporative flux was used:

$$(10.6) \qquad E/\overset{*}{E} = \rho(g/T)^{1/2} \left| \frac{\partial \theta}{\partial z} \right|^{1/2} (z - d)^2 \frac{\partial q}{\partial z}$$

During forced convection, a logarithmic regression analysis yielded

$$(10.7) \qquad \overset{*}{E} = 0.210 |\text{Ri}|^{-0.5}$$

During free convection,

$$(10.8) \qquad \overset{*}{E} = 1.402$$

The transition from forced to free convection occurred at about $\text{Ri} = -0.025$ but the transition point was not sharp.

During inversions, the quantity $\overset{*}{E}$ decreased rapidly with increasing Ri and there was a large experimental scatter of points; the scatter was probably caused by the intermittent presence of gravity waves and drainage winds. For Ri between 0.01 and 0.05,

(10.9) $\overset{*}{E} = 0.174|\text{Ri}|^{-0.5}$

In all these regressions, the values of Ri and $\overset{*}{E}$ referred to a height of about 66 cm above the zero-plane displacement.

10.5. The Eddy Correlation Method for Measuring Evaporation

From equation (8.24) it follows that

(10.10) $Q_E = LE = L\,\overline{\rho q' w'} \simeq L\,\rho\,\overline{q' w'}$

Equation (10.10) provides a fundamental method for measuring evaporation when $\bar{w} = 0$. Although the instrumental difficulties are formidable, good estimates may be obtained provided that the averaging time is at least 5 or 10 min and provided that a hot-wire anemometer is used to measure $\rho w'$. The evapotron [121, 124] measures vertical fluxes of both heat and water vapor. Taylor [135] has emphasized the importance of matching the time constants of the various sensors.

10.6. Some Practical Considerations

It has frequently been suggested that experimental measurements should be made as close to the ground as possible; at these low levels the profiles are supposed to converge to the logarithmic form. The advice is certainly to be followed in the application of equation (10.3). However, radiative flux divergence is likely to be greatest near the ground, which suggests that in the use of equation (10.4), there is merit in remaining a meter or so above the surface.

Equation (10.4) is preferred to (10.3) because:

1. Experimental estimates of u_* are not reliable.
2. An assumption about the value of the ratio K_H/K_E is probably preferable to one about the value of K_m/K_E.

Figure 31 compares estimates of evapotranspiration at Davis, California [136] obtained from a weighing lysimeter and from equations (10.4) and (1.1), assuming $K_H/K_E = 1$. The results are typical of the good agreement that was found except when there was advection or during periods of strong instability (when the Bowen ratio overestimated evaporation by 50 to 100%).

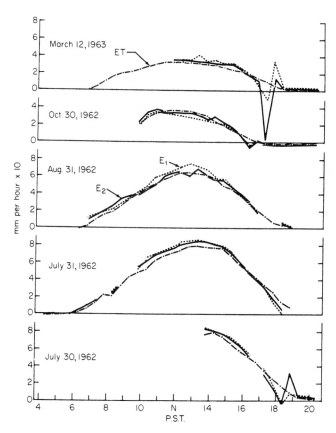

FIG. 31. Lysimeter evapotranspiration at Davis, California versus energy balance estimates of evaporation, E_1 (using profile data in the 25–50-cm layer—dotted line), and E_2 (using profile data in the 50–100-cm layer—solid line) [136].

There is sometimes a need to determine vertical fluxes of carbon dioxide, ozone, or tritium. A recommended procedure is to assume similarity of the diffusivity of the substance with K_E (rather than K_m or K_H) and to calculate the flux from vertical gradients for the same layer as was used in determining K_E.

10.7. Measurement of Temperature and Water Vapor Fluctuations

Temperature fluctuations can be measured with a fine-wire resistance thermometer or thermocouple. Zwang [126] used a platinum resistance thermometer, 20 μ in diameter with a time constant of 0.01 sec. With a sensor that small, shielding and ventilation are no longer necessary. There has also been recent interest in the sonic thermometer (e.g., ref. [137]), which integrates the temperature fluctuations over a path-length of from 10 to 100 cm.

Water vapor fluctuations may be measured with a fast response wet-bulb resistance thermometer or thermocouple. Taylor and Dyer [121] used a nickel resistance thermometer (0.0005 in. in diameter with a time constant of 0.3 sec). Alternatively, Elagina [138] has developed an optical device based on the absorption of radiation by water vapor at 1.38 μ, using a path-length of 1 meter.

11. Wind Flow around Obstacles

11.1. The Surface of the Earth

A homogeneous "infinite plane" is the exception rather than the rule in nature. The countryside is often hilly while over level ground there are usually hedges, fences, or buildings. It is therefore of great practical importance to consider the effect of obstacles on the wind flow.

Because of the large number of irregular geometric shapes, the problem is not capable of mathematical characterization except for a few simple cases such as flows around spheres or cylinders. The wind tunnel has consequently been widely used to investigate scale models of bluff objects. In the strong-wind case with sufficient turbulent mixing to ensure adiabatic conditions, scale-model results have generally been reproducible in the atmosphere, e.g., Woodruff and Zingg [139].

11.2. Wind Flow around a Cylinder

Flow around cylinders will serve as a useful example. In the nonviscous case, fluid particles are accelerated over the upstream half of the obstacle and decelerated thereafter. Work done by pressure forces is converted into kinetic energy between A and B (Fig. 32) while the reverse is true from B to C.

In the viscous case, the fluid particle dissipates some of its kinetic energy between A and B. Consequently, it cannot fully overcome the adverse pressure differential between B and C (Fig. 32); a vortex develops at C and there is *boundary layer separation*.

Figure 32 also includes the flow pattern that develops a few seconds later. At regular intervals a vortex detaches itself from the wake of the cylinder and moves downstream (with a velocity less than that of the undisturbed fluid). In due course a regular series of vortices, known as a *Kármán vortex street*, moves downstream, with alternating clockwise and counterclockwise rotation. The frequency n of vortex shedding has been found experimentally to depend upon the *Strouhal number*, St $= nD/u$, where D is the diameter of the cylinder. For fully turbulent flow, St appears to be constant with a value near 0.21.

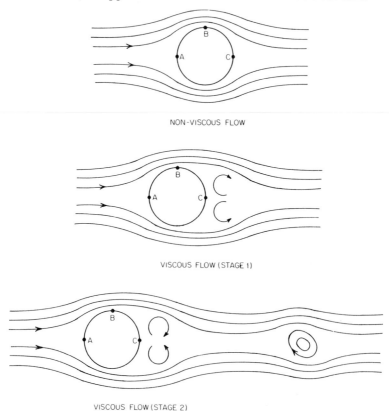

FIG. 32. Schematic representation of nonviscous and viscous flows around a cylinder.

The occurrence of vortices has been related experimentally to values of the Reynolds number of from about 60 to 5000. At lower Reynolds numbers the wake is laminar while above Re $= 5000$, there is complete turbulent mixing.

Turbulence in the wake of a cylinder displays intermittency. For a sufficiently long averaging period, however, the character of the mean flow and of the turbulent fluctuations can be predicted from physical models. A comprehensive treatment has been given by Hinze [84].

11.3. Wind Flow around Irregular Objects

Except in the case of very light winds, turbulent wakes develop around buildings, hedges, and other obstructions to the wind. The fluxes of momentum, heat and water vapor in this turbulent zone have not been studied in any detail. This is because of the dominating influence of the geometry of the obstacle, which makes it impossible to obtain reproducible results.

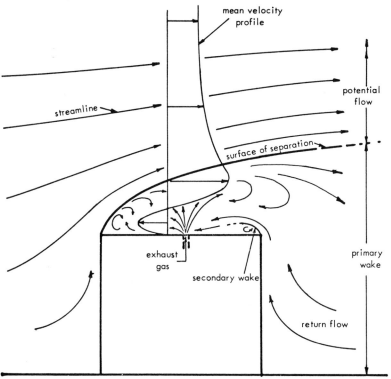

FIG. 33. Typical flow pattern around a cube with one face normal to the wind [140].

Figure 33 [140] is a schematic representation of typical wind tunnel behavior around a building. There is a *surface of separation*, above which the flow is accelerated. Below the surface there is a turbulent wake in which counter-flows develop. The velocity profile indicates

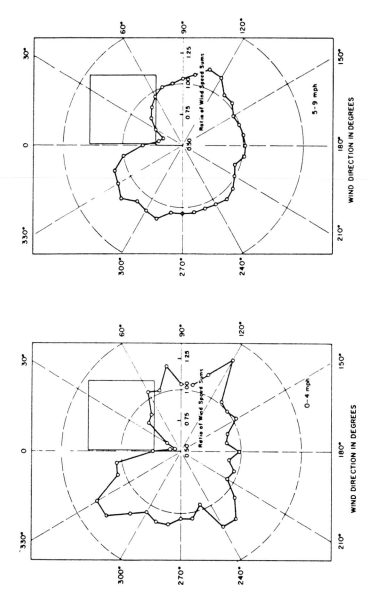

Fig. 34. Ratio of tower anemometer speeds to pole anemometer speeds as a function of wind direction measured on tower (relative size of tower and boom indicated in northeast sector [141]).

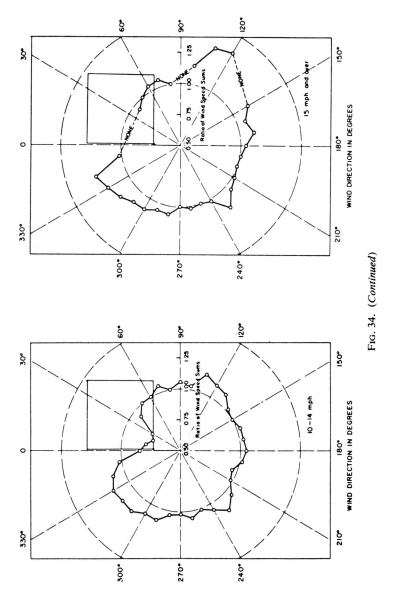

FIG. 34. (*Continued*)

that there is a layer of air above the roof of the building in which the mean flow is reversed.

Figure 33 represents average conditions over a few minutes. From second to second, the surface of separation "bounces" up and down; in addition, there is shedding of Kármán vortices. A further complication not present in the wind tunnel occurs in the atmosphere; because of the presence of low frequency fluctuations upstream, the angle of attack of the undisturbed flow is continually changing.

11.4. The Energy Balance of an Enclosed Area

In view of the behavior illustrated in Figs. 32 and 33, how is it possible to estimate the energy balance of a fenced field or of a clearing in a forest? The flux equations assume steady state conditions and a uniform horizontal flow of infinite extent. There are therefore many uncertainties.

One approach is to locate the sensors at the center of the field and as close to the ground as possible. The assumption is then frequently made that edge effects in some sense cancel, permitting extrapolation to the entire area. It is more usual, however, to deliberately select open country with long uniform fetches. The micrometeorologist who chooses to take measurements in an enclosed field can never be certain of the reproducibility of his results in areas having slightly different dimensions. For this reason, published micrometeorological data are biased toward uniform terrain despite the importance of the study of uneven surfaces. In a city, for example, there are urgent requirements for engineering estimates of wind loadings on tall buildings and for the prediction of the transport and diffusion of pollution.

11.5. The Effect of a Tower on Wind Measurements

Moses and Daubek [141] have studied the disturbing effect of an open tower on wind speeds at Argonne National Laboratory in the United States. Winds were measured at the 18.75-ft level of a 150-ft fire-lookout tower having a cross section of 22 sq ft at that level; the anemometer was mounted on a boom projecting 9 ft outward. The hourly winds were compared with 5000 simultaneous observations from a second anemometer positioned on top of a pole well away from the tower but at the same height above ground.

The wind shadow effect is illustrated in Fig. 34. When the wind blew through the tower before reaching the anemometer, there was a substantial reduction in speeds, the effect being greatest with light winds. When the wind was from the northwest or southeast, on the other hand, there was an increase in speeds.

Rider [142] has studied the effect of a 4.5-meter mast (diameter 2.5 cm) on cup anemometer measurements. The instruments were positioned at six levels on side-arms 25 cm in length. Two of the anemometers were on the opposite side of the mast to the other four, and it was found that smooth wind profiles could not be drawn from the data. A wind tunnel test was then undertaken using a constant speed of 8 meters/sec. The sequence of positions occupied by the anemometer is given in Fig. 35, and the results are shown in Table X.

The 90° arc over which the variation was smallest is shown in Fig. 35. Rider recommends that anemometers be mounted to the

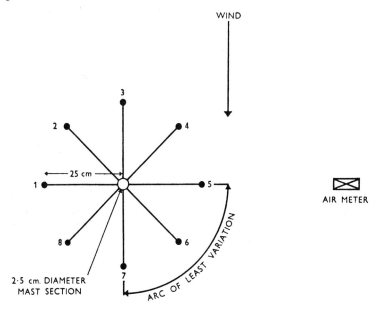

FIG. 35. Sequence of positions occupied by an anemometer in the wind tunnel during a series of 10-min runs [142] (by permission of Her Britannic Majesty's Stationery Office).

right of the mast trailing 45° behind the normal to the expected wind direction.

Finally, some recent and as yet unpublished studies were made at Suffield, Canada by O. Johnson in 1964. A cup anemometer was mounted at a height of 10 ft on a 1-in pipe. Another anemometer was 30 ft away (cross wind) at the same height and at the end of a boom extending upwind from a triangular tower of 3-ft sides. The boom was adjustable so that the anemometer could be positioned as indicated in Table XI. The wind speeds were averaged over a number of 30 to 60 min periods when mean speeds ranged from 6.8 to 20.2 mph.

TABLE X. Variation of anemometer indication according to mounting position in a wind tunnel [142].

Position	1	2	3	4	5	6	7	8
Ratio of air meter to anemometer wind speeds	1.028	1.009	1.008	1.026	1.000	0.996	1.002	0.995

TABLE XI. Percentage reduction of tower-mounted anemometer over control anemometer.

Distance of anemometer from tower, ft	$1\frac{1}{2}$	2	3	5	6	8	10
Percentage reduction of wind speed	5.2	4.7	2.35	1.5	1.5	0.8	0.75

These data may be compared with a rule of thumb often quoted by Gill [143] that an anemometer should be mounted outward from a television tower at a distance greater than the diameter of the tower at that height.

12. Transitional Zones and States

12.1. Introduction

Transitional zones are regions in which the flow is readjusting itself to a new set of boundary conditions in space. The mathematical models assume that some kind of discrete change takes place along a line $x = 0$, that the flow is at right angles to the line, and that steady state conditions exist upwind of the discontinuity. The problems then fall into four categories:

1. The effect of a barrier such as a fence or a hedge. The approach is rather negative in the sense that the main interest lies not in the study of profiles within the turbulent wake, but rather in the determination of the fetch required for the profiles to return to their undisturbed upwind state.
2. The effect of a discrete change of roughness z_0. What kind of readjustment, for example, takes place when air flows from a short grass to a long grass surface?
3. The effect of a discrete change in surface temperature and moisture, e.g., advection of air from land to lake with a resulting redistribution of the available energy within the two media. In this connection, a change in surface roughness also takes place but is frequently ignored.
4. The extension of point 3 to cases in which the properties of the underlying surface are continually changing, e.g., advection of air across a body of water in which there is a gradient of surface water isotherms at right angles to the flow.

Transitional states are caused by a change in the available energy input commencing at some time $t = 0$. There may be a change in net radiation Q_n or perhaps a change in moisture when a field is irrigated. The transition may be continuous (the diurnal cycle of Q_n under clear skies) or may be discrete (the appearance of a cloud shadow).

The atmosphere is always in a state of transition on some scale or other. However, the main body of theory uses simplified models in which only a few of the parameters are permitted to vary.

12.2. The Fetch Required to Achieve Steady State Conditions Downwind from an Obstacle

The effect of a fence at right angles to the wind is known qualitatively. Recovery to steady state conditions appears to occur in an exponential fashion. Hence, it is not possible to find a discrete downwind distance where the disturbed flow ends. Instead, it is necessary to consider downwind distances in which 95, 99, or 99.9% recovery is achieved. In this connection, the mean wind speed returns to normal more quickly than does the turbulence. Hence, the appearance of smooth wind profiles at some downwind distance x does not necessarily imply that the turbulent statistics have returned to their upwind values.

Experimental results and resulting rules of thumb are conflicting. First, this is partly because of differences in upwind turbulence; the way in which the disturbed wake interacts with the pre-existing turbulence is not understood. Second, the height of the obstacle H must have some effect, although most results are reported in terms of the dimensionless parameter x/H instead of in terms of x. It is true that flows around cylinders and spheres in homogeneous turbulence preserve similarity for different diameters. However, the situation must be quite different in a shear zone.

Priestley [93] suggests as a guide that meaningful profiles will be found up to a height of one-twentieth the distance traversed by the wind over a uniform surface. However, Brooks [144] believes that tree interference effects may be important at distances of 50 tree heights if an accuracy of 3% is required. Inoue *et al.* [145] recommend that measurements in a small field should be made only to a height of 1/100 of the fetch and that the field should be at least as wide as the distance from the sensors to the upwind edge.

The opinions quoted above may be used as a guide but it is evident that more research needs to be done in this important aspect of micrometeorology.

12.3. The Effect of a Discrete Change in Roughness

Elliott [146], Taylor [147], and Panofsky and Townsend [148] have considered the adiabatic case in which the roughness changes discretely at $x = 0$ from $z_0{}'$ to $z_0{}''(z_0{}' > z_0{}'')$. It has been found in the wind tunnel that there is a transition zone between flows in which the friction velocity is not constant. However, at a downwind distance of $x_1 (x_1 \simeq 7\frac{1}{2}z_0')$, a new boundary layer begins to form at the ground and gradually thickens. This has been called the *internal boundary layer* by Elliott.

The simplifying assumptions differ in the three papers and it cannot be said that a general solution has been found. However, some of the results are qualitatively similar. If h is the height of the internal boundary layer at downwind distance x, Elliott finds that

$$h \propto x^{0.8} \qquad \text{for } x/z_0'' > 10^3$$

Panofsky and Townsend broaden the assumptions but find also that the height of the interface varies as the fetch to the 0.8 power.

When the effect of stability is introduced, the equations become intractable. However, some qualitative arguments by Elliott suggest that the interface height h is increased in lapse conditions and decreased during inversions, although the effect is relatively small.

It would appear that vertical profiles will show no discontinuities and that the shearing stress will be independent of height, provided that the height is less than one-twentieth the fetch. However, Panofsky and Townsend emphasize that a horizontal gradient of shearing stress may continue for long distances, i.e., $\partial u_* / \partial x \neq 0$. In the real atmosphere, where the internal boundary layer must grow to a thickness of 500 meters before complete equilibrium is reached with the geostrophic wind, the horizontal variation in friction velocity may be significant for large distances.

Some pertinent experimental data have been presented by Kutzbach [149]. Five hundred bushel baskets were placed on ice-covered Lake Mendota, Wisconsin in the arrangement indicated schematically in Fig. 36. The upwind fetch across the lake was at least 2000 meters from the control anemometer, which was at a height of 340 cm. The results from one particular experiment are given in Fig. 37, for which the "specific area" of baskets, i.e., the reciprocal of the area density of baskets, was 2 sq meters. The seven 10-min wind profiles were obtained over a period of $2\frac{1}{2}$ hr and the control wind varied slightly (see inserted legend in Fig. 37). Nevertheless, a plot of the ratio of profile wind to control wind shows a reproducible trend for fetches (R) of 20 and 40 meters. At a fetch of 48 meters, the wind speed is still decreasing, which indicates that equilibrium conditions have not yet been reached. The experiments have subsequently been extended to

12. *Transitional Zones and States*

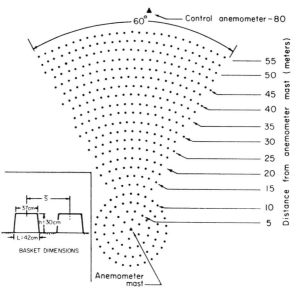

Surface roughness modification study
bushel basket experiment – 1961 Fef.15

FIG. 36. The Wisconsin bushel basket experiment on the ice of Lake Mendota [149].

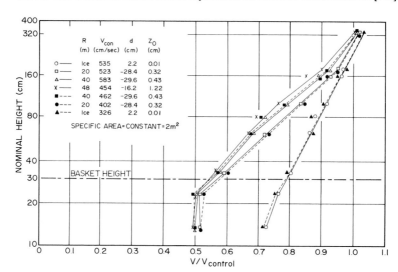

FIG. 37. Wind profiles associated with different fetches (R) downwind from a discrete change in roughness on Lake Mendota, 1513–1749 CST, Feb. 19, 1961 obtained by introducing bushel baskets as indicated in Fig. 36 [149].

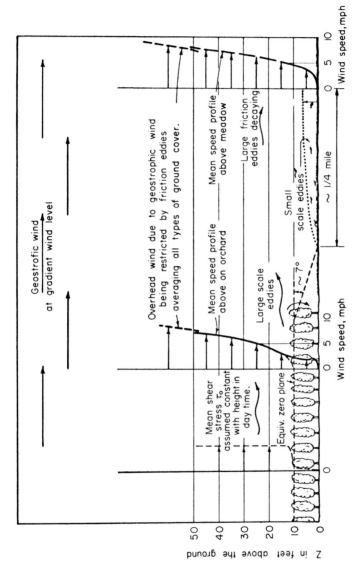

FIG. 38. Schematic representation of wind passing from extensive orchard to open field [144].

include an array of 228 Christmas trees artificially "planted" in the ice-cover of Lake Mendota [150].

These theoretical and experimental studies emphasize the caution that should be exercised in the interpretation of profiles obtained from "average" locations. Wind profiles from tall towers may be influenced by terrain many kilometers away, and it is quite possible that experimental values of z_0 and d will vary with wind direction.

Figure 38 [144] is a vivid example of changes in mean wind flow as it passes from orchard to meadow.

12.4. Advection

When the wind flow is at right angles to a discontinuity in temperature or humidity, or to a gradient of these, a "box" model argument [similar to that used in developing equation (5.6)] will show that

$$(12.1) \qquad \frac{\partial T}{\partial t} + \bar{u}\frac{\partial T}{\partial x} = \frac{\partial}{\partial x}\left(K_{Hx}\frac{\partial T}{\partial x}\right) + \frac{\partial}{\partial z}\left(K_{Hz}\frac{\partial T}{\partial z}\right)$$

It is generally assumed that:

1. Steady state conditions exist, i.e., $\partial T/\partial t = 0$.
2. Diffusion in the forward direction is negligible in comparison with forward transport by the mean wind. Thus it is permissible to neglect the term $\partial/\partial x[K_{Hx}(\partial T/\partial x)]$.

With these restrictions, equation (12.1) becomes

$$(12.2) \qquad \bar{u}\frac{\partial T}{\partial x} = \frac{\partial}{\partial z}\left(K_H\frac{\partial T}{\partial z}\right)$$

Similarly,

$$(12.3) \qquad \bar{u}\frac{\partial q}{\partial x} = \frac{\partial}{\partial z}\left(K_E\frac{\partial q}{\partial z}\right)$$

It will be noted that if $\partial T/\partial x = 0$ and $\partial q/\partial x = 0$, the equations reduce to equations (9.4) and (10.1).

In order to solve (12.2) and (12.3), it is necessary to know $\bar{u}$ and K in terms of z. In addition, the boundary conditions must be carefully specified.

Many solutions have been offered (they have been referred to as "mathematical extravaganzas" by F. Gifford), but the results have been generally disappointing. A review of the methods has been given by Sutton [56].

One difficulty is that there are interactions between horizontal temperature and moisture gradients. A frequently used boundary condition in equation (12.3) when dry air blows across a moist surface is that

(12.4)
$$\lim_{z \to 0} C = \text{constant} = C_s \qquad (x > 0)$$

where C_s is the surface concentration. Since the surface concentration is dependent upon temperature, and since the partition of available energy in equation (1.1) is changing downwind, it is unlikely that equation (12.4) will be satisfied.

One approach is to use an empirical modification of equation (12.4). Portman and Brock [151] have solved equation (12.3) on an analog computer, using as a boundary condition

(12.5)
$$\lim_{z \to 0} C = \alpha(1 - e^{-\beta x}) \qquad (x > 0)$$

where α and β are empirical constants.

Philip [152] has solved equations (12.2) and (12.3) simultaneously, using equation (1.1) as a boundary condition. The power law profiles for wind and diffusivity, equations (7.21) and (7.6), are applied, with the further assumption that the exponent p has a value of $1/7$.

Rider and Philip [153] consider the implications of Philip's theory for a specific case in which the upwind values of the variables are chosen arbitrarily. The results are given in Table XII.

TABLE XII. Numerical example of advective effects [153].

Position relative to discontinuity	Q_H, 10^{-3} ly/sec	Q_E, 10^{-3} ly/sec	Bowen ratio
Upwind	4.50	0.50	9.00
1 meter downwind	−1.21	6.21	−0.19
10 meters downwind	0.08	4.92	0.02
100 meters downwind	1.08	3.92	0.28
1 km downwind	1.85	3.15	0.59
10 km downwind	2.45	2.55	0.96
100 km downwind	2.91	2.09	1.39

The appearance of a negative value of Q_H with a consequent inversion in temperature is to be noted at a distance of 1 meter downwind. This kind of behavior has been verified qualitatively by Dyer and Pruitt

[154]. The same authors clearly establish the existence of flux divergence of Q_H and Q_E between the surface and a height of 4 meters over an irrigated field at Davis, California. As the summer proceeds and the surrounding countryside dries out, the magnitude of flux divergence increases.

In this connection, an interesting point has been raised by Swinbank [112] and by Sheppard [155]. The buoyancy term, $gQ_H/\rho c_p T = -u_*^3/k\Omega$ (see Section 9.6), is strictly true only in dry air or when the vertical gradient of mixing ratio is zero. In other cases, the buoyancy term should be written

$$\frac{g}{\rho}\left(\frac{Q_H}{c_p T} + \frac{3}{5}\frac{Q_E}{L}\right)$$

For $Q_E = Q_H$, the correction is negligible. However, for $Q_E = 14Q_H$ the two terms in brackets are of about the same magnitude. An adiabatic lapse rate might not then represent neutral equilibrium over an irrigated field in a desert climate; instead, there would be upward buoyancy.

Rider *et al.* [156] have tested Philip's model at Canberra Airport, Australia during periods when the air flow was across 350 meters of tarmac ($z_0 = 2 \times 10^{-3}$ cm) before reaching a grass surface ($z_0 = 1.4 \times 10^{-1}$ cm). The observed humidity changes were in good agreement with Philip's model but the temperature changes were only about half of those predicted. The error is attributed to the change in roughness, which is not included in the model.

Dyer [157] has considered the rate of adjustment of Q_H and of the vertical temperature gradient under the assumption that completely dry air moves over an area where the evaporation is everywhere constant and that the lapse rate is near the adiabatic. Using Philip's solutions, Dyer obtains estimates (Table XIII) of the fetch required for 90% adjustment.

The purpose of the exercise was to provide engineering estimates for the location of sensors downwind from a leading edge. Although the underlying assumptions are not likely to hold in any real case, Table XIII does give realistic order-of-magnitude readjustment values.

Miyake [158] studied the modification of wind and temperature at Pt. Barrow, Alaska when cold air (about $-19°C$) moved over open sea water ($-1.7°C$). The height of the internal boundary layer was clearly defined in the temperature profiles (about 30, 150 cm at fetches of 10, 20 meters, respectively). The wind profiles cannot be interpreted so readily.

Finally, the work of Priestley [93] on large-scale air mass modification should be mentioned. Among other things, he considers the advection of cold air over different types of surface. At the leading

TABLE XIII. Values of fetch and height
for 90% adjustment [157].

Height, meters	Fetch, meters
0.5	70
1	170
2	420
5	1350
10	3300
20	8100
50	26,500

edge of the cold outbreak, the heat gain by the air is dependent upon K_H rather than upon K_G. However, after an hour or so, further modification is largely determined by the heat capacity of the ground. In the case of dry sand, there is very little additional heating of the air whereas modification continues for many hours over a water surface.

12.5. Transitional States

Transitional states involve a net loss or gain of energy to the earth-atmosphere system; it is required to determine how this loss or gain is shared among the terms of equation (1.1).

The fundamental equation is again (12.1), but in this case the partial derivatives with respect to x are assumed to be zero. As a result,

$$(12.6) \qquad \frac{\partial T}{\partial t} = \frac{\partial}{\partial z} \left(K_H \frac{\partial T}{\partial z} \right)$$

Similarly,

$$(12.7) \qquad \frac{\partial q}{\partial t} = \frac{\partial}{\partial z} \left(K_E \frac{\partial q}{\partial z} \right)$$

Sutton [56] and Priestley [93] have reviewed the elegant solutions that have been obtained for these equations. The choice of initial and boundary conditions is often artificial, particularly in the specification of a diffusivity, but the chief failing appears to be in the independent solution of the equation without reference to equation (1.1). This is possible through the adoption of some prescribed periodic form for the temperature at $z = 0$. The assumption predetermines the general form of the final solution.

If the net radiation drops by 50%, then the loss must be shared in some unknown way by Q_H, Q_E, and Q_G. An analogous approach to that used by Philip in the solution of equations (12.2) and (12.3) has considerable merit, therefore. Development along these lines is proceeding by Estoque [159]. With the help of a digital computer, Estoque has been able to simulate many important characteristics of the boundary layer. He recognizes that his model is still an oversimplification of real conditions and suggests that future studies should include soil moisture effects through equation (5.8).

Finally, Dyer [157] has considered a limiting case of a completely dry atmosphere and a horizontally uniform site in which the evaporation suddenly changes from zero to some constant value. The atmosphere is assumed to be adiabatic with a power law for the diffusivity. The analysis also applies to any other conservative diffusing entity, e.g., a sudden change of surface temperature due to the appearance of a cloud bank. The time required for the profiles to reach 90% adjustment to the new regime is given in Table XIV, and it

TABLE XIV. Time required to reach 90% level of adjustment after a sudden change in surface conditions [157].

Height, meters	Time, min
0.5	0.45
1	0.99
2	2.18
5	6.20
10	13.8
20	30.4
50	86.0

is seen that more than an hour is required for 90% readjustment at a height of 50 meters. Dyer concludes that: " In experiments intended to assess the variation of transfer coefficients (or some similar parameter) with stability, one is frequently concerned with assessing slight departures from the neutral situation. In the author's view, disparity between a number of such experiments may well be due to failure to satisfy the basic requirements, namely, a horizontally uniform site and steady state conditions.''

13. Atmospheric Pollution

13.1. The Meteorological Problem

The essential problem can be put very simply. Suppose that pollution is being emitted from a point in space at a height h above the ground. Given the emission strength Q (mass per unit time), it is required to calculate the resulting concentrations (mass per unit volume) at any point in space and for various averaging times. The interest extends from the "lungful of air" concentration to the lifetime dosage.

The introduction of line or multiple sources adds nothing fundamentally new because the effects are additive. However, the volume source problem, e.g., the dispersion of a cloud of particles, is quite different. The treatment will not be included in this chapter but an excellent discussion can be found in Pasquill [86].

Early (and some recent) attempts at estimating diffusion from a continuous point source in steady state conditions commenced with the differential equation of flux, similar to equation (12.1):

$$(13.1) \qquad \bar{u}\frac{\partial \chi}{\partial x} = \frac{\partial}{\partial x}\left(Kx\frac{\partial \chi}{\partial x}\right) + \frac{\partial}{\partial y}\left(Ky\frac{\partial \chi}{\partial y}\right) + \frac{\partial}{\partial z}\left(Kz\frac{\partial \chi}{\partial z}\right)$$

where χ is the concentration of pollution. It is assumed in most models that the pollution does not undergo significant physical or chemical change during its residence in the atmosphere. In the simplest kind of theory, the particles are also assumed to be weightless; if they have a fall velocity, another term must be added to equation (13.1).

The diffusivity in the x-direction K_x may be neglected in comparison with forward transport by the wind. The equation can then be solved, given appropriate boundary conditions and functional forms for $\bar{u}$, K_y, and K_z. A similarity argument is often used to equate K_z to the momentum diffusivity K_m. However, there is no mathematical or intuitive way of selecting a form for K_y. Boundary layer theory yields no direct information about lateral diffusivity; as a result, the only case that can be solved with confidence is diffusion from a line source at right angles to the wind direction. Equation (13.1) is therefore the basis for many elegant solutions that are rarely used operationally.

13.2. A Diffusion Model from Probability Theory

The *normal* probability distribution is given by a bell-shaped curve whose equation is

$$(13.2) \qquad E = \frac{1}{(2\pi)^{1/2}s} \exp\left\{ - \frac{(\bar{x} - x')^2}{2s^2} \right\}$$

where E is the probability density function, x' is a fluctuation about $\bar{x}$, and s^2 is the variance of the fluctuations. The factor $(2\pi)^{1/2}$ is included so that the total area under the curve from $x = -\infty$ to $x = +\infty$ will be equal to unity.

In two dimensions, when the distribution is *binormal*, the equation becomes:

$$(13.3) \qquad E = \frac{1}{2\pi s_x s_y} \exp\left\{ - \frac{1}{2} \left[\frac{(\bar{x} - x')^2}{s_x^2} + \frac{(\bar{y} - y')^2}{s_y^2} \right] \right\}$$

This is a basic working tool in diffusion theory.

Let there be a continuous point source Q of gas or weightless particles at a height h. Let the concentration at any point (x, y, z) be χ. It is assumed that diffusion in the forward direction may be neglected in comparison with the wind, and that the particles diffuse in a random fashion in the y- and z-directions. Thus, any vertical cross section through the plume will have a binormal distribution of particle positions, with variances s_y^2 and s_z^2. An equation similar to (13.3) will then apply.

$$\bar{u}\chi/Q = \frac{1}{2\pi s_y s_z} \exp - \frac{1}{2} \left\{ \frac{y^2}{s_y^2} + \frac{(z - h)^2}{s_z^2} \right\} .$$

It is noted that the probability density function E of equation (13.3) has been replaced by $\bar{u}\chi/Q$ to preserve dimensions and to take account of the fact that the mean wind stretches the plume forward with resulting dilution. This equation is not yet complete because vertical diffusion is bounded by the surface of the earth. It is assumed [56] that the ground acts as a perfect reflector, and that there is a *mirror image* source at $z = -h$ as well as one at $z = +h$, as indicated in Fig. 39.

The final form of the equation is then:

(13.4)

$$\bar{u}\chi/Q = \frac{1}{2\pi s_y s_z} \exp(-y^2/2s_y^2)\left[\exp{-\frac{(z-h)^2}{2s_z^2}} + \exp{-\frac{(z+h)^2}{2s_z^2}}\right]$$

For ground-level concentrations ($z = 0$), or for a ground-level source ($h = 0$), or both, equation (13.4) can be simplified somewhat.

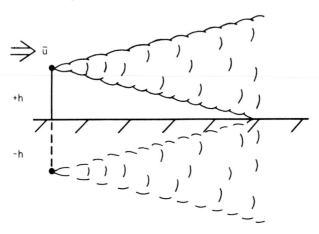

FIG. 39. Sutton's mirror image model for diffusion from a continuous point source.

The model assumes that the mean wind is constant in space and time (including no vertical wind shear) and that the particle positions in the plume form *normal* distributions, although the lateral spread may be different from that in the vertical direction. The approximate nature of resulting estimates can therefore be appreciated. It is well known, for example, that a cross section of the plume at any instant rarely has a *normal* probability particle distribution. However, the ground-level concentration is usually an average over a period of 3 min or longer, and it is a plausible assumption that the smoothed cross sections will then have *normal* distributions.

Another inconsistency occurs near the origin where the model predicts that there is a small but finite probability that a particle will have moved a large but finite distance within a few seconds. This objection applies also to solutions of equations (13.1) and (12.1). Monin [160] recommends use of the so-called telegraph equation which has hyperbolic rather than parabolic solutions. However, the question is hardly of operational importance in the application of an empirical model.

Since a plume expands with increasing distance from the source, s_y and s_z must be functions of x. A number of relations have been proposed either empirically (e.g., Cramer [161]), in which case the reproducibility at other locations is open to question, or through the application of Taylor's theorem (described below), which is set in the Lagrangian reference frame.

13.3. Taylor's Theorem

Suppose that a weightless particle is released into a homogeneous atmosphere at some point $(0, 0, 0)$. After T seconds the particle will have moved to some new position (x_1, y_1, z_1). Because of the random nature of turbulence, the point (x_1, y_1, z_1) is indeterminate.

Assume that the turbulence is stationary in the statistical sense and let the experiment be repeated many times. The various positions of the particle T seconds after release may be designated as (x_1, y_1, z_1), (x_2, y_2, z_2), ... , (x_n, y_n, z_n). These points constitute a probability ensemble and will have a center of gravity $(\bar{x}, \bar{y}, \bar{z})$ and three variances, σ_x^2, σ_y^2, σ_z^2, each having the dimensions of (length)^2. If desired, the discrete particle may be replaced mathematically by a continuum of fluid elements. Hence, the model is applicable to gaseous as well as to particulate releases.

Taylor [162] derived a theorem that applies to the above experiment. The proof is given in [56, 86]. Using the notation of Chapter 8,

$$(13.5) \qquad \sigma_y^2 = 2 \overline{v'^2} \int_0^T \int_0^t R(\xi) \, d\xi \, dt$$

where v' is the instantaneous particle velocity. There is a similar expression for σ_z. The theorem was derived originally for one-dimensional turbulence but it may be applied to the projection on a single axis of three-dimensional turbulence when particle dispersion is jointly, as well as separately, normally distributed.

The limiting cases of equation (13.5) are readily derived. For very small values of t, $R(\xi)$ is approximately equal to unity.

$$(13.6) \qquad \sigma_y^2 = \overline{v'^2} \, t^2$$

For very large values of t, using equation (8.19),

$$(13.7) \qquad \sigma_y^2 = 2 \overline{v'^2} \, l_2 t$$

Unfortunately, there is no way of deciding the limiting times for which equations (13.6) and (13.7) are applicable.

For intermediate times it is reasonable to assume that

(13.8) $$\sigma_y^2 \propto t^{2-n}$$

where the exponent n lies between the values of 0 and 1. Sutton's widely used model [56] is based on the assumption that there is a usefully broad range of times for which the index n may be considered as constant.

Taylor's theorem applies to a hypothetical experiment in which all particle positions correspond to equal travel times. In the atmospheric problem of calculating diffusion from a point source, on the other hand, the quantities of interest are s_y and s_z [see equation (13.4)], which are derived from distributions of particles having varying atmospheric travel times. The two reference frames are related by the *ergodic principle*, time averages taken along the path of one particle are equivalent to space averages taken over a number of particles. This implies that $\overline{v'^2}$, $\overline{w'^2}$ are the same in the Lagrangian and Eulerian systems and that $s_y^2 = \sigma_y^2$, $s_z^2 = \sigma_z^2$; the autocorrelation functions are, however, different. The ergodic assumption is reasonable in the case of steady state conditions and homogeneous turbulence. In the atmosphere, unfortunately, these conditions are rarely present although tacitly assumed in experimental studies.

13.4. The Pasquill Diffusion Model

It may be helpful first to visualize qualitatively the behavior of a plume in a turbulent wind over a homogeneous surface. At any distance x from the source, there is an upper limit to the size of eddy that can contribute to diffusion. The very large eddies cause a meandering of the plume center-line but do not change s_y or s_z. Furthermore, as x increases and the plume expands, the size of eddy capable of making a contribution to diffusion increases.

An additional factor to be considered is that downwind concentrations are usually measured over a discrete time interval, say, 10 min. In this case, the meandering of the plume over a typical 10-min period must be included in diffusion calculations; s_y and s_z are larger than for instantaneous sampling of the plume, although there is again an upper limit to the size of eddy that can contribute to diffusion.

It therefore seems intuitively plausible that certain wave bands in the turbulence spectra must be related in some way to the values of s_y^2 and s_z^2, the appropriate wave bands being dependent upon the distance from the source and the sampling time.

These ideas can be stated in a quantitative way by introducing equations (8.10) to (8.12). The Lagrangian variances, σ_y^2 and σ_z^2, may be considered as the integrals of spectral functions when appropriate weighting functions are added to correct for the effect of discrete smoothing and sampling times. Provided that the sampling time is greater than the smoothing time, and using the assumption contained in equation (8.26), Pasquill was able to convert Taylor's theorem to the form:

$$(13.9) \qquad \sigma_y^2 = s_y^2 = \overline{v'^2}\, T^2 \int_0^\infty F_E(n) \left[\frac{\sin \pi n T/\beta}{\pi n T/\beta} \right]^2 dn$$

where $F_E(n)$ is the Eulerian spectral function and β is the ratio of Lagrangian to Eulerian time scales.

Pasquill's model does not require the calculation of spectra. Presmoothing of a wind fluctuation record over moving intervals of time t/β before variances are computed will provide a solution to equation (13.9); t is the plume travel time.

Finally, since it is easier to measure azimuth and elevation angles with a bidirectional vane than it is to obtain v' and w', a conversion is made to

$$(13.10) \qquad (\sigma_A)_{T,\,t/\beta} = \tan^{-1}(s_y/x)$$

$$(13.11) \qquad (\sigma_E)_{T,\,t/\beta} = \tan^{-1}(s_z/x)$$

where σ_A, σ_E are the standard deviations of azimuth and elevation angles, sampled over an interval of time T, the discrete time interval over which the mean plume dimensions are required. The other subscript t/β indicates that the bivane traces are smoothed over moving intervals of time t/β before standard deviations are computed. Devices for electronically determining the standard deviations σ_A, σ_E have been developed [163, 164] and are commercially available.

Pasquill's model is recommended for use with an elevated source during steady state conditions over homogeneous terrain within a kilometer or so of the source.

13.5. The Effect of Lapse Rate on Plume Behavior

The behavior of a plume of smoke or gas is determined by the mean wind and its turbulent fluctuations. During superadiabatic lapse rates, there is considerable low frequency energy in the spectrum of turbulence; dilution rates will then be high except near the source where there will be *meandering* (in the lateral direction) and *looping* (in the vertical) but no large expansion of the instantaneous plume.

When inversions are present, on the other hand, the levels of turbulence and the mean wind are low. Dilution rates are small and the plume can remain visible for many kilometers. Most of the serious smog incidents have been associated with an inversion that persisted for several days. No existing diffusion equation is capable of predicting dilution rates when the wind is light.

Because turbulent fluctuations are difficult to analyze, lapse rate has become a useful empirical index in air pollution studies. Television towers may be equipped with temperature sensors (see Fig. 16, for example), and the data are used in the statistical interpretation of urban air quality measurements and in chimney height design.

Another qualitative index is the daily *maximum mixing depth* [165], the height to which there is vigorous daytime vertical mixing of pollution. Mean values at Pittsburgh, for example, range from 340 meters in January to 1510 meters in May.

13.6. Effective Stack Height

The physical models of diffusion up to this point have assumed that the emission has no exit velocity and that it is at the same temperature and density as its environment. This is not generally the case. Gas often leaves a chimney with a considerable exit velocity and elevated temperature. It is therefore necessary to define an *effective stack height*, i.e., a height at which the plume becomes approximately horizontal. This new height is then used for the quantity h in equation (13.4).

Unfortunately, the determination of effective stack height is one of the most notoriously difficult problems in diffusion. Plume rise depends upon the exit velocity and temperature of the gas, as well as upon the lapse rate, wind speed, and turbulence structure of the environment. The plume becomes level more quickly when there are strong winds or an intense inversion. However, the specific way in which momentum and heat are transferred from the jet is a question of much speculation and has resulted in a number of conflicting theories.

One special case for which there appears to be agreement is when there are calm winds and an inversion. Spurr [166] has produced a nomogram that yields a good estimate of the heat that must be added to an effluent in order that buoyancy forces will carry the plume up through the top of an inversion.

When a wind is blowing, no model appears to yield clearly superior results. This is in part a result of the scarcity of experimental data. Moses and Strom [167] tested 6 formulas at the Argonne National Laboratory and found that no one formula was outstanding. The temperature excess was only slight so that the results are not likely to be applicable to the stack of a large smelter. Lucas *et al.* [168] have measured plume rises from two large power stations in England. Their data can be represented by the expression:

$$(13.12) \qquad\qquad Z_{max} = \alpha Q^{1/4}/\bar{u}$$

where Z_{max} is the plume rise in feet, Q is the rate of emission of heat in megawatts, and $\bar{u}$ is the wind speed in feet per second. The " constant " α is affected by lapse rate and topography; average values of 4900 and 6200 were found for the two power stations. Rauch [169] has photographed plume rises from two chimneys (height 125 and 75 meters; exit velocities 4 to 12 and 2 to 6 meters/sec; gas temperatures 120°C and 250 to 400°C, respectively). Rauch favors equation (13.12) but finds an average value of α of 2070.

When it is necessary to calculate diffusion from a chimney, the best advice that can be given is to use several plume rise formulas. A probable upper and lower limit for effective stack height can then be established, from which it will be possible to " bracket " the resulting ground-level concentrations.

13.7. Aerodynamic Downwash around a Building

A large building is an obstacle to the wind. This creates aerodynamic downwash on the leeward side when the wind speed exceeds a particular critical value (see Fig. 33, for example). The wake turbulence dilutes the plume but it is possible to have occasional instantaneous concentrations very close to source strength.

Because the geometry of the building is important, it is customary to study scale models in the wind tunnel. By trial and error, a plant design may be found which minimizes downwash.

The calculation of ground-level concentrations in the wake of a building is a difficult but operationally important problem. There are few experimental data and, in addition, the reproducibility of results to other buildings is questionable. In November 1962 releases of a tracer (uranine dye) were made [170] from the top of the chimney (within the heated gases) of the National Research Council Central Heating Plant in Ottawa, Canada. Ground-level 20-min concentrations were measured with six samplers located a few hundred feet to the rear of the building. Bidirectional vanes were mounted both on the windward side and to the lee of the building. Some typical results are given in Fig. 40 and Table XV. The effect of the obstacle on the

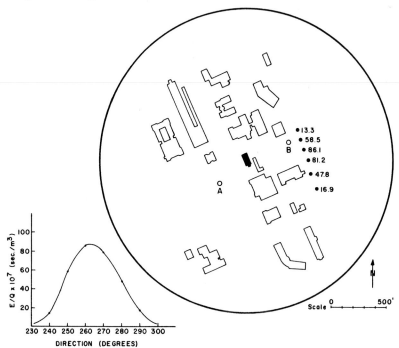

Fig. 40. National Research Council, Ottawa. Uranine dye was released from the stack of the heating plant (center of diagram in black). Height of release was 70 ft above ground. Ground-level concentrations were measured at the indicated positions during 1605–1625 EST, Nov. 8, 1962 yielding the distribution of χ/Q shown in the graph at the lower left. Bivanes were at points A (82 ft above ground) and B (33 ft above ground). Mean wind speeds at A were 263/6.5 mph at 20 ft and 275/7.9 mph at 80 ft [170].

turbulence is illustrated, as well as the smooth pattern of ground-level concentrations that emerges when a 20-min sampling time is used.

Important theoretical and experimental studies are in progress by Halitsky [140, 171]. He generalizes equation (13.4) to the form:

$$(13.13) \qquad\qquad \bar{u}\chi/Q = K/A$$

where K is a dilution coefficient and A is a reference area such as the area of the front side of a bluff object. The coefficient K combines in one parameter all the meteorological turbulence statistics. Halitsky

TABLE XV. Bidirectional vane standard deviations ahead of (and at a height of 82 ft) and in the wake (and at a height of 33 ft) of a heating plant in Ottawa, Canada, 1605–1625 EST, Nov. 8, 1962 (20 min sampling period).

Smoothing time sec	Azimuth angle, deg		Elevation angle deg	
	Ahead	In wake	Ahead	In wake
5	11.2	18.7	4.4	10.3
10	10.7	17.3	3.8	8.1
25	9.7	15.3	3.0	5.5
50	8.9	14.1	2.4	3.9
100	8.3	13.1	1.7	3.0

believes that although K will vary with position and sampling time, it may be independent of scale; i.e., wind tunnel and atmospheric data may yield similar values of K.

13.8. Transitional Zones and States

When air pollution data from a city are averaged by hours of the day, it is usually found that minimum values occur in early afternoon when horizontal and vertical ventilation are greatest. (An exception occurs in the case of oxidants and ozone, whose production depends upon photochemical reactions and sunlight.) There are two maxima, one an hour or so after sunrise and another in the evening. The morning peak is partly caused by the increase in emissions at that time of day and is partly due to a *Hewson fumigation* [51] associated with the transition from inversion to lapse conditions.

The evening pollution maximum is not so clearly understood. The radiation inversion does not grow uniformly during the night but thickens as a series of discrete pulses. A *low-level jet stream* [172] tends to develop at the top of the inversion. With each readjustment of inversion depth, the wind maximum disappears and pollution is transferred up and down. This may account for the evening pollution peak.

A fumigation also occurs when a lake or sea breeze blows inland during the spring and summer. Convection builds upward with distance inland in the transitional zone. Stack gases moving inland in the stable advection layer may therefore fumigate to ground level at the point where the plume intersects the top of the new internal boundary layer.

Hewson [173] has given a simple formula for calculating fumigation concentrations, based on a geometric argument:

$$(13.14) \qquad \bar{u}\chi/Q = \frac{36}{\pi h(x_2 + x_1)}$$

where χ is the average value of the ground concentration occurring over a distance interval from x_1 to x_2 downwind from the source (for the quarter to half-hour period of highest concentrations), and h is effective stack height. The model assumes that the plume widens horizontally over a $5°$ angle from the source but otherwise is relatively free of restrictive assumptions. Equation (13.14) gives engineering order-of-magnitude estimates of fumigation concentrations.

When calculating diffusion in a valley or over other nonhomogeneous surfaces, there is considerable merit in assuming uniform rates of dilution within some simple geometrical volume.

14. The Air over Bare Ground

14.1. Models and Reality

Chapters 7 to 12 described idealized models of micrometeorological phenomena. It must be evident that the theory is far from complete, even within the framework of the simplifying assumptions that were made about the nature of the boundary conditions.

Despite these limitations, the theory does provide guidance in the interpretation of data obtained from the real environment of forests, lakes, snow surfaces, and cities, for example. Because of the economic importance of micrometeorology in such matters as world food production, water conservation, forest regeneration, and air pollution, this account would be incomplete without describing local weather patterns that prevail over various natural surfaces. By including both the physical models and the applications, it is hoped that the gap may be narrowed and in some ways bridged between theory and application.

14.2. The Energy Balance of a Dry Surface

The simplest situation that could be imagined is air flow over a flat desert. Evaporation can be neglected in that case and the energy balance equation (1.1), reduces to

$$(14.1) \qquad Q_n = Q_G + Q_H$$

Measurements in such an environment should yield important information about the form of Q_H and the variation of K_H with height and Richardson flux number. It is indeed surprising that so little effort has been made in this direction. A well-designed study would include the measurement of net radiation, radiative flux divergence, profiles of wind and temperature, soil heat flux, and the covariances $\overline{u'w'}$ and $\overline{w'T'}$.

The emphasis in arid zone research has been along lines of more practical importance to the economy, namely, studies of temperature extremes, of the formation of dew (an important factor in the water balance of dry regions), and of the effectiveness of irrigation.

The diurnal variation of temperature over a dry bare surface is extreme, largely because of the absence of evaporational daytime cooling and condensational heating at night, which exert modifying influences. In addition, heat flows into and out of the ground Q_G are decreased. It is well known, for example, that frost is more likely over a dry than over a moist surface.

Factors that prevent even higher daytime temperatures are:

1. The high albedo of dry sand (see Table II).
2. The high temperature of the radiating surface and the resulting large value of $Q_{L\uparrow}$ [see equation (3.1)].

A comparison of the energy balance in the desert and in an irrigated oasis is given in Figs. 41 and 42 [9]. The daytime net radiation Q_N is

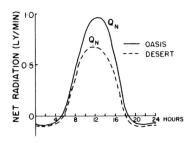

FIG. 41. The net radiation balance in an oasis and in a desert [9].

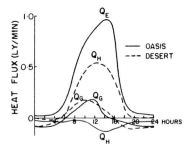

FIG. 42. The energy balance in an oasis and in a desert [9].

larger in the oasis because of the decreased albedo and greater absorption of short-wave radiation by the surface. Evaporational cooling is absent in the desert (Fig. 42) but is of major importance in the oasis: in fact, it results in negative values of Q_H throughout the day and a temperature inversion.

The subject of dew has always been of interest in arid zone research. In this connection, Monteith [174] distinguishes between:

1. *Dew fall*: condensation of water vapor from the atmosphere.
2. *Distillation*: condensation of water vapor moving upward from the soil.
3. *Guttation*: the release of liquid from certain parts of the leaves.

An excellent review of the dew mechanisms and of their relative importance in the water balance of crops has been given [27]. The only component that represents a clear gain to the local water economy is dew fall, and this at times may be a small fraction of the total, particularly under very stable lapse rates with light winds. Fortunately, dew fall can be measured directly with a lysimeter; distillation and guttation will not be recorded by that instrument. In principle, dew fall can also be deduced indirectly from the energy balance equation (1.1) or by the eddy correlation method, equation (10.10). However, these methods are of doubtful value during strong inversions.

Dew fall appears to be of economic importance in arid regions near large bodies of water such as along the coast of the Mediterranean. Farther inland, dew is of little significance except perhaps to the natural desert plants which transpire slowly and for which dew represents a clear gain in their water balance.

Another dry surface of both theoretical and practical interest is concrete or pavement. There is no local evaporation but presumably a plentiful supply of water vapor is being advected from neighboring grass or wooded areas. The engineer is interested in maximum daytime surface temperatures as well as frost penetration depths in winter. The micrometeorologist, on the other hand, views the area as an interesting transitional zone. In addition to the question of readjustment of profiles, he would like to find a sufficiently long fetch of concrete so that equation (14.1) could be applied under steady state conditions to determine the form of K_H. Some pioneer studies have been undertaken in the United States [175] and Australia [153, 156] but much more remains to be done.

For engineering purposes, the best available estimate of pavement surface temperature is obtained from equation (3.2), i.e., by assuming that the surface radiates as a blackbody. The question of frost penetration rates is considered in Chapter 15.

14.3. The Energy Balance of Moist Ground

When soil moisture is not limiting and the terrain is uniform, the standard energy balance methods outlined in Chapters 9 and 10 are available. For qualitative estimates of evaporation from moist soil (and from vegetation), Penman has suggested on a number of occasions that *monthly averages* can be obtained with an error of not more than 20% by assuming that $Q_E = Q_N$, i.e., $Q_G + Q_H = 0$. The assumption *must not*, of course, be used to estimate an instantaneous or an hourly value of Q_E.

As the soil dries, evaporation no longer continues at the potential rate. It is still possible to obtain the sum $(Q_H + Q_E)$ by the energy balance method, and it is clear that Q_H will increase as Q_E decreases for the same environmental conditions. The question that has not yet been answered satisfactorily is whether the water vapor diffusivity K_E is affected by an insufficient supply of soil moisture. This may perhaps be a "nonsense" question similar to that concerning the value of the thermal diffusivity K_H when the lapse rate is adiabatic and the heat flux is zero. In any event, the partition of energy between Q_H and Q_E is possible by using the lysimeter or indirectly through the eddy correlation method, equation (10.10).

14.4. The Effect of Fences and Hedges

The requirements for fetch suggested in Chapter 12 are not met in many cases. In present agricultural applications, an observing site is chosen over uniform ground; the results are then extrapolated frequently to an entire township or region. Although beset with a number of theoretical objections, the estimates may often be qualitatively useful indices, exhibiting the correct diurnal and seasonal variations, and providing a statistical separation of windy-dry and calm-humid days. Quite satisfactory correlations may therefore be obtained with plant growth or flower production. The accuracy with which evaporation and sensible heat estimates must be made depends upon the ultimate use to which the estimates will be put.

15. The Air over Snow and Ice Surfaces

15.1. Introduction

Energy balances of snow and ice surfaces are unique. Snow is almost a perfect blackbody in the long-wave part of the spectrum but it also has a very high short-wave albedo. The surface roughness is small, reducing turbulence and vertical heat transfer. Latent heat exchange includes the additional effects of sublimation and of heat of fusion; for example, when the temperature is at the freezing point, part of the available energy is used in melting snow rather than in evaporation. Finally, the thermal conductivity of snow is much lower than that of soil.

There are many practical reasons for studying energy transfers at snow and ice surfaces. For example, there is a need to predict ice thickness for navigation and frost penetration for the construction and road-building industries. Hydrologists are interested in spring run-off from snow fields while agriculturists are concerned with the moisture storage of a blanket of snow. The glaciologist interprets the growth and decay of glaciers in terms of energy balance relations. Furthermore, the Arctic is no longer a great white wasteland; expanding industrial activity has led to an urgent requirement for micrometeorological information of polar regions. Still another example is the profound effect that regional snow-cover anomalies must have on macroscale weather patterns. Although the interactions are not yet understood, they undoubtedly influence the general circulation of the atmosphere.

15.2. The Radiation Balance of Snow and Ice Surfaces

A snow surface is often called a *radiative sink*. This is because of strong reflection of solar energy (high albedo) and strong radiation

133

of terrestrial energy (high emissivity). In addition, the heat conductivity of snow is low; hence, surface radiative heat losses are not replaced quickly by heat fluxes from below.

Solar radiation not lost by reflection at the surface penetrates to a depth of up to 1 meter in snow [13] and 10 meters in ice [176]. The attenuation can be approximated by Beer's law, equation (2.3), but values of the extinction coefficient are quite variable, depending upon the physical properties (such as compaction) of snow and ice. There is also a variation with wavelength; for example, Liljequist [177] finds a greater penetration of blue than of red light in a snow cover at Maudheim, Antarctica.

A detailed study of extinction coefficients in ice has been reported by Lyubomirova [178]. The attenuation was almost independent of wavelength in the 0.43–0.63-μ band but increased by 10% in the 0.63–0.73-μ band. There was also a variation with the type of ice and with depth. Average values for samples obtained from the Baksan and Azau rivers in the Soviet Union are given in Table XVI.

TABLE XVI. Attenuation of solar radiation in river ice [178].

Ice thickness, cm	Penetration of solar radiation in % of radiation that has entered	Extinction coefficient, a, cm^{-1}
1.9	40	0.480
3.3	25.5	0.410
5.3	24.5	0.280
8.8	17.7	0.196
16.0	11.4	0.136
27.0	8.0	0.094
46.0	3.8	0.073

Table XVI represents typical values but it must be recognized that there are wide variations dependent upon the distribution and size of the air bubbles in the ice.

When radiation instruments are placed above snow and ice, part of the measured reflected energy has come from below the surface. Liljequist [177] reports that diffuse upward blue light originating within snow was 40% of the downward flux.

Long-wave radiation cannot penetrate ice. The exchange of terrestrial energy therefore takes place in the upper millimeter or so of the surface. Light fluffy snow, on the other hand, permits some long-wave exchange in the upper 3 cm [179]. The relative attenuations of short- and long-wave radiation in snow are shown schematically in Fig. 43, from which it may be inferred that a "greenhouse" effect

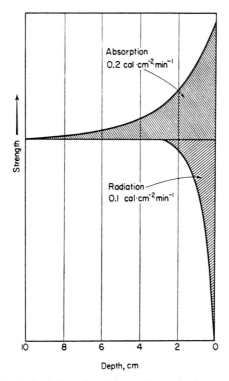

FIG. 43. Theoretical calculation of positive and negative heat sources within snow due to absorption and radiation [179].

must occur below the 3-cm level. In some cases radiative heat gains at 3–10 cm may be sufficient to cause melting while the surface remains frozen.

15.3. Heat Flux and Heat Storage within Snow and Ice

Temperature profiles within snow and ice depend upon:

1. The energy balance at the upper surface.
2. The energy balance at the lower soil or water interface.

3. Conductivity and heat capacity.
4. The penetration of solar radiation.
5. In the case of loosely packed snow, convective and latent heat exchange within the snow.

During hours of darkness, there is a sharp temperature gradient in the upper part of the snow cover, largely because of its low thermal conductivity. An ice surface does not become as cold as a snow surface because of greater heat transfer through the medium.

During daylight hours in summer, snow and ice quickly come to a uniform temperature of 0°C throughout because of strong insolation and the percolation of water to deeper levels. The quantity Q_G then becomes negligible, permitting some simplification of equation (1.1).

When a vertical temperature gradient exists in loosely packed snow, the possibility of convective as well as conductive heat transfer must be considered. Murcray and Echols [180] found a gradient of 0.59°C/cm in winter snow at Fairbanks, Alaska. However, the fact that large temperature differences persist for weeks implies that the effect of convection cannot be large. Yen [181] has calculated theoretically that a vertical flow of air of 1 cm/sec through snow increases its effective thermal conductivity from 0.0014 to 0.0020 in centimeter-gram-second units.

The saturated vapor pressure over snow depends upon temperature. Hence, there is a flow of water vapor from snow of higher temperature to that of lower temperature. The latent heat flux has been calculated by Yen [182] to increase the effective thermal conductivity by 7.5% in still air and by 19% when the vertical air flow is 1 cm/sec.

Calculations of the energy balance over bare ground were based on the fact that the radiation stream was entirely intercepted by the upper surface. Heat flow through the ground Q_G was by conduction and could be determined from a series of vertical temperature profiles. In the case of snow, ice, and water surfaces, however, solar radiation penetrates the media; temperature gradients are then partly due to radiative flux divergence and are not indicative of the heat flow reaching the upper surface. The artificial assumption is often made that the radiation stream terminates at $z = 0$, but this leads to a substantial remainder term in equation (1.1).

To overcome the difficulty, and for other reasons, it is sometimes useful to calculate the energy balance of a *volume* of snow or ice; equation (1.1) is then no longer applicable.

The winter heat balance study of Lake Mendota, Wisconsin is an illustrative example [183]. Since the lake was covered with both ice and snow, separate determinations were required for each medium:

(15.1)   *Ice* $Q_S = Q_{AS} + Q_{IW} + Q_{IS}$

(15.2)   *Snow* $Q_S = Q_{AS} + Q_{IS} + Q_{L\downarrow} - Q_{L\uparrow} \pm Q_H \pm Q_E$

When there is no snow above the ice, equation (15.1) becomes

(15.3) $Q_S = Q_{AS} + Q_{IW} + Q_{L\downarrow} - Q_{L\uparrow} \pm Q_H \pm Q_E$

In these equations, Q_S is the rate of storage of sensible and latent heat in the snow or ice as indicated by the rise or fall in temperature through the medium or by the change in volume due to increasing ice thickness. The short-wave radiation absorbed by the medium per unit time is Q_{AS}. The heat exchange rates through the ice-water and ice-snow interfaces are Q_{IW} and Q_{IS}, respectively.

The storage of sensible heat is given by

(15.4) $$Q_S = \rho c \int_0^z \frac{\partial T}{\partial t} \, dz$$

In many midwinter cases, temperature changes with time are negligible, and the important heat storage effect is latent. The *latent heat storage* of ice is given by the product of the rate of change of ice thickness, the density, and the latent heat of freezing; the latent heat storage of snow is given by the product of the rate of change of the weight of snow per unit area and the latent heat of freezing.

The absorption of solar radiation by the medium is the most uncertain term in the balance equations. On the basis of a literature review, Scott and Ragotzkie [183] used extinction coefficients in Beer's law of 0.25 cm^{-1} and 0.05 cm^{-1} for snow and "bubbly" ice, respectively, in midwinter.

The conduction terms Q_{IW} and Q_{IS} may be determined from temperature profiles, although some error is introduced by the penetration of solar radiation.

In theory, equations (15.1) to (15.3) could be used to predict the growth of ice or frost penetration. In practice, however, the individual terms cannot be determined with sufficient accuracy to provide reasonable estimates of Q_S. Present methods are therefore largely empirical, and depend upon climatological factors such as the persistence of cold weather, the depth of snow, and the wind speed.

15.4. Wind Profiles over Snow and Ice

When lapse rate is adiabatic, the wind structure over snow and ice is often represented by a logarithmic profile, equation (7.11). In some instances, a zero-plane displacement d is required to take account of the wavelike ridges of snow, called *sastrugi*. Values of the roughness length z_0 range from 0.0001 to 0.07 cm; Liljequist [177] suggested that z_0 increases with increasing wind speed but Dalrymple *et al.* [60] found no such effect.

During inversions, Liljequist was able to match wind profiles in the 60-cm to 10-meter layer at Maudheim, Antarctica to a " log + linear " law, similar in form to equation (7.20) but developed empirically and independently of Monin and Obukhov [59].

15.5. Temperature Profiles and Vertical Heat Transfer

When winds are strong, the lapse rate is adiabatic over snow and ice. In contrast with conditions over bare soil, however, daytime insolation rarely creates a superadiabatic lapse rate; the surface temperature cannot exceed 0°C and excess energy is used in melting the snow. Dalrymple *et al.* [60] found a few cases of negative Richardson numbers during the polar night, which they attributed to long-wave radiation exchange with warm air aloft.

Inversions are common (see Fig. 17, for example) and may persist for several weeks during the polar night; they are ultimately destroyed when strong winds develop. Liljequist [184] found a marked reduction in inversion intensity when the wind speed at the 10-meter level reached 8 to 9 meters/sec.

There are reports of "anomalous" profiles of both temperature and wind. The minimum temperature may occur a few centimeters above the surface as it does over grass; the cause is radiative flux divergence, shallow slope winds (see Chapter 20), or ice crystal sublimation [185].

The South Pole data of Dalrymple *et al.* [60] are unique in the sense that they contain many cases of intense inversion with measurable winds in the 50- to 800-cm layer. At other locations the wind usually becomes so light that it is not possible to calculate a Richardson number from equation (9.6). Monin [116] predicted that in strong stability, ϕ/z could no longer be a function of height and that the wind must then increase linearly with height (see Chapter 9).

It follows from equation (7.14) that the Deacon number, De, is zero when the wind profile is linear.

Figure 44 displays the variation of De (called β_v by Dalrymple) with Ri. The solid line is a solution of equation (9.14) using empirical values for the constants. It can be seen that the Deacon number did not fall to zero with increasing stability; on the contrary, De reached a minimum value of about 0.25 for Ri = 0.15, and then began to increase again.

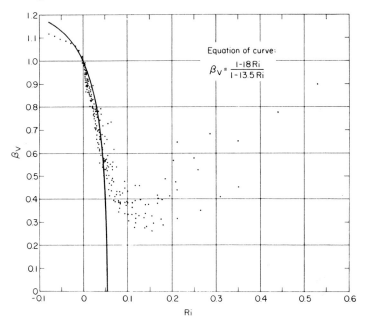

Equation of curve:

$$\beta_V = \frac{1 - 18\,Ri}{1 - 13.5\,Ri}$$

FIG. 44. Variation of Deacon number for wind with Richardson number at the South Pole [60].

The behavior of DE [equation (7.15)] (called β_θ by Dalrymple) with Ri is given in Fig. 45. The Deacon number for temperature became negative with strong stability and showed considerable scatter; similarity of wind and temperature profiles was therefore not a good assumption during intense inversions. Dalrymple *et al.* believe that these results are consistent with a model of flux divergence of both heat and momentum.

Not many estimates exist of the heat flux Q_H over snow and ice surfaces. Kazansky [176] has suggested an average daily downward

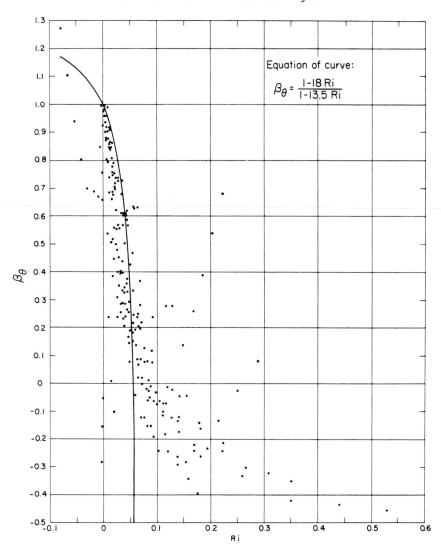

FIG. 45. Variation of Deacon number for temperature with Richardson number at the South Pole [60].

transfer of 0.43 ly/min over the Fedchenko glacier in the U.S.S.R. in summer; the average net radiation Q_N was 0.455 ly/min. Liljequist [184] proposed an empirical relation for winter inversion conditions at Maudheim:

(15.5) $$Q_H = -0.0058\bar{u}_{10} \quad \text{(ly/min)}$$

where $\bar{u}_{10}$ is the standard 10-meter level anemometer wind.

During very cold dry weather, the value of Q_E is small and can sometimes be neglected. The quantity Q_H can then be found from equation (1.1) when the sun is below the horizon.

15.6. Humidity Profiles and Latent Heat Transfer

Because of experimental difficulties in measuring vapor pressures at low temperatures, profiles of vapor pressure or of mixing ratio are practically nonexistent during winter conditions. It is usually assumed that sublimation to water vapor is negligible but that frost or ice crystal formation may heat the surface and the air slightly.

Even for temperatures between $-10°C$ and $0°C$, there are few data. However, Uchida [186] undertook a remarkable study of wind, temperature, and vapor pressure profiles within 10 mm of a snow surface, using a thermistor-anemometer, a thermistor-thermometer, and an electric frost-point meter. The average flux of water vapor was calculated to be upward at 0.177 gm/m² min in the afternoon and downward at 0.004 gm/m² min in the evening of March 19, 1960 on the Shiga Plateau, Japan.

When snow and ice are melting at $0°C$, some of the available latent heat is used to melt the cover (80 cal are required to melt 1 gm of snow) while some is used for evaporation or sublimation (597 cal are required to evaporate 1 gm of water while 597 + 80 cal are required to sublime 1 gm of snow). For computational purposes, sublimation may be neglected. If H is the total water-equivalent rate of ablation from a snow or ice surface [187]:

(15.6) $$Q_E = 597E + 80(H - E)$$

The quantity $(H - E)$ is the melting rate in centimeters of water per unit area per unit time. It is not at all obvious, given Q_E, how the available energy is distributed between E and $(H - E)$. Supplementary measurements of ablation rates are therefore required, or alternatively, estimates of evaporation obtained from vertical profiles as well as estimates of Q_E as the remainder term in equation (1.1). A difficulty with the latter method is, of course, that equation (1.1) does not apply when the sun is shining and radiant energy is penetrating below the interface.

Some typical results for the Blue Glacier in the United States are given in Fig. 46 [188]. The glacier was at a temperature of 0°C and ground heat flux Q_G was assumed to be zero.

Kazansky [176] determined the heat exchange of the Fedchenko glacier in summer. In the upper reaches of the glacier, specific humidity decreased with height in the 0.25- to 4-meter layer; some of the

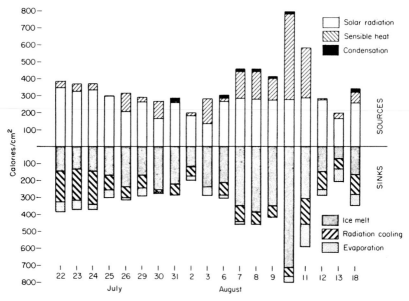

FIG. 46. Daily components of the energy balance on the Blue Glacier, in the state of Washington [188].

available latent heat was therefore being used in evaporation. In the central and lower areas of the glacier, on the other hand, specific humidity increased with height and the flow of water vapor was directed downward. As a result, latent heat of condensation was available for melting the ice (averaging 0.05 ly/min over a day). Kazansky's study illustrates the fact that when warm air is advected across a snow surface, the air may release its latent heat to accelerate the melting.

16. The Energy Balance of a Plant Cover

16.1. The Air over a Short Grass Surface

There has been a number of careful determinations of the energy balance over short grass. The most widely quoted are those of Rider [67], Swinbank [98], and Lettau and Davidson [26].

Rider's results are typical. A circular area of approximately 250 meters in diameter was kept mown to a grass length of 2 to 3 cm at Cardington, England. Profiles of temperature, humidity, and wind were obtained within the layer from 15 to 200 cm. The radiation components, soil temperatures, and the surface drag were also measured. Some observations are given in Table XVII, the Richardson number being for the 37.5-cm elevation.

TABLE XVII. The energy balance over short grass (ly/min) [67].

	Observation number					
	31	33	35	44	46	48
Time GMT	1100	1410	1118	1815	2100	1815
$Ri \times 10^3$	-18	-23	-77	4	4	30
$Q_T - Q_R$	0.568	0.329	0.459	—	—	—
$Q_{L\downarrow} - Q_{L\uparrow}$	-0.168	-0.142	-0.144	-0.122	-0.118	-0.116
Q_N	0.400	0.187	0.315	-0.122	-0.118	-0.116
Q_G	0.046	0.015	0.043	-0.055	-0.035	-0.082
Q_H	0.150	0.058	0.188	-0.126	-0.142	-0.049
Q_E	0.204	0.114	0.084	0.059	0.059	0.015
K_m	1360	730	—	990	1670	280
K_H	1540	1190	2610	1940	2250	590
K_E	1450	980	930	1650	2060	—

Observations 31 and 33 show a similar pattern; the percentage of available energy used in transpiration was 51 and 61%, respectively.

In No. 35, on the other hand, only 27% was directed toward Q_E and the relative magnitude of Q_H went up accordingly. The larger negative Richardson number suggests a regime of free convection.

After dark, the absolute values of Q_H and Q_E are much smaller. There is slight transpiration while the sensible heat flux is directed downward.

16.2. The Energy Balance of a Leaf

Gates [189] has measured leaf temperatures at Boulder, Colorado using a radiometer and applying equation (3.2). He suggests that although the emissivity of green leaves is 96 to 98%, the assumption contained in equation (3.2) of 100% emissivity leads to an error of only $\pm 0.2°C$ in his data.

When a leaf was perpendicular to the sun's rays, its upper surface temperature was quite variable, averaging 8 to 10°C above the air temperature but on one occasion being 20°C higher. The lower surface was about 1.5°C cooler than the upper surface. There appeared to be no systematic variation with size of leaf or species. However, temperature differences decreased with increasing wind speed.

Shaded leaves (and all leaves on overcast days) averaged 0.1 to 2.4°C below air temperature for six species, the lower surface again being cooler than the upper.

In order to estimate transpiration rates, it is necessary to determine the convective heat flux component Q_H. Gates used empirical wind tunnel equations relating to heat transfer from flat plates and cylinders. For an oak leaf containing no stomata on its upper surface, Gates then estimated that the transpiration heat loss was 0.583 ly/min, i.e., a transpiration rate of 0.00099 gm/cm² min. This is the loss from a single leaf at right angles to the solar beam, and the average for an entire plant or tree would, of course, be much less.

In the case of a pine needle (cylinder), the convective transfer was greater than for a deciduous leaf. For a needle temperature of 3°C above air temperature, Gates calculated a value of 0.160 ly/min for Q_E, yielding a transpiration rate less than one-quarter of that for deciduous leaves in the same environment.

Inoue [190] doubts whether such an approach can ever lead to reliable quantitative estimates. There is a gradient of air temperature over a leaf and it is not possible to specify precisely an air-leaf temperature difference. Furthermore, leaves flutter in the wind and convective heat flux will not be analogous to that from a rigid flat plate in the wind tunnel. Nevertheless, data on upper and lower surface temperatures of shaded and unshaded leaves would be helpful in the interpretation of concurrent micrometeorological observations taken within and above a plant cover. Gates' results suggest that transpiration rates, and values of Q_E, can be significantly different for various kinds of vegetation even within the same environment (see also refs. [191, 192]).[3]

16.3. The Radiation Balance of a Plant Cover

The radiation balance above vegetation can be measured in the usual way. It has been found that reflectivity varies with wavelength [193] and with solar altitude [194] because of absorption of specific wavelengths by plants and because of multiple reflections from beneath the canopy.

An important consideration is the penetration of solar radiation into the plant cover. As in the case of a snow or ice surface, it is not possible to assume that the radiation stream terminates at some specific level. Attenuation will depend upon the altitude of the sun (producing diurnal and seasonal variations) and the height and density of the vegetation. The density is usually specified by the *leaf area index*, LAI.

$$(16.1) \qquad \qquad \text{LAI} = \int_0^h A(z)\, dz$$

where $A(z)\, dz$ is the leaf area between z and $(z + dz)$ per unit area of ground, and h is the height of the vegetation canopy. It is often assumed that there are no leaves below $z = h/4$.

The radiation balance at various levels within a plant community is not well documented experimentally largely because of horizontal inhomogeneities, and it is likely that space-mean averages will be

---

[3] K. R. Knoerr and L. W. Gay [(1965). Tree leaf energy balance. *Ecology*, **46**, 17–24)] have recently measured leaf transpiration directly with a *potometer*, a calibrated small glass tube attached with sealing wax to the petiole of a freshly cut leaf. If leaves in potometers have the same temperatures as attached leaves in the same environment, it is assumed that transpiration rates are similar. For sweetgum and tulip poplar leaves, the latent heat exchange rates were estimated to be about 0.15 and 0.05 ly/min in sun and shade, respectively, in a subsequent study to be published.

required at different heights and for various leaf area indices and solar altitudes. Nevertheless, a good empirical assumption appears to be given by Beer's law, equation (2.3), provided that plant leaves are not concentrated in one thin layer. Allen *et al.* [195] have measured net radiation in a field of corn; the average height of the upper leaves (not the tassels) was 300 cm. The resulting radiation profiles are given in Fig. 47 for a clear day, September 10, 1961, at Ithaca, New York.

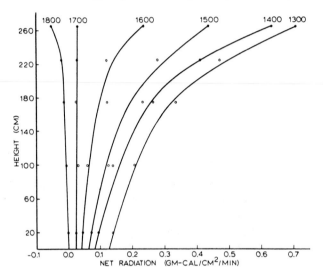

FIG. 47. Net radiation in a field of corn, Sept. 10, 1961 [195].

There is a scatter of points but a Beer's law relation seems appropriate.

$$(16.2) \qquad (Q_N)_z = (Q_N)_h \exp - [a(h - z)]$$

Another way of expressing this relation has been suggested by Uchijima [196].

$$(16.3) \qquad (Q_N)_z = (Q_N)_h \exp -(a\mathrm{LAI}_z)$$

where the extinction coefficient a has a value of about 0.6 and the leaf area index is for the layer from z to h.

In addition to the flux divergence within vegetation, botanists are interested in the changing spectral distribution. Geiger [13] emphasizes that forest shade is different from that found in the shadow of a building. Specific wavelengths are absorbed by the upper leaves of a plant. Although much of the remainder may be reflected down to lower levels, it may lack the energy necessary for photosynthesis.

16.4. Profiles within a Plant Cover

In general, the warmest daytime temperatures are found part way down into the plant cover. An example is given in Fig. 48 [197] for

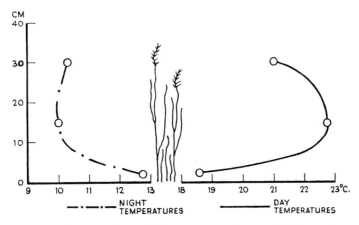

FIG. 48. Night and day temperatures in grass [197].

a field of long grass. There was an inversion of temperature from the ground to a height of 10 cm at a time when a profile over short turf would show a lapse condition. In autumn, because of the angle of the sun, the inversion layer becomes thicker.

The situation is reversed at night. A lapse condition may exist up to a height of 10 cm (Fig. 48) with an inversion above. These are average conditions, and many small-scale fluctuations or "ripples" may occur [198]. In addition, dew formation will release latent heat and reduce the vertical temperature gradients.

Two exceptions should be noted. In the first place, when the plants are far enough apart to permit direct penetration of the sun's rays to the ground, temperature profiles may resemble those to be found over a bare surface. Second, for certain crops having very high transpiration rates, the latent cooling may cause a temperature inversion to form above the plants.

The humidity within a plant cover is high, and generally increases toward the ground. However, if transpiration is concentrated within a discrete vertical layer of leaves, the slope of the mixing ratio profile may change within that layer.

Wind speed decreases downward within a crop. A typical example is given in Fig. 49 [197] for a silage crop, from which it may be seen that

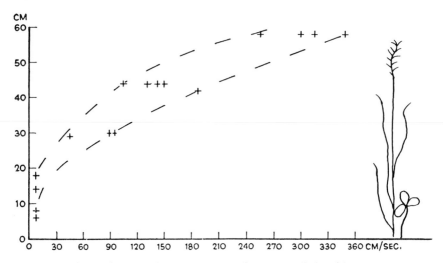

FIG. 49. Wind profiles in a silage crop. Broken lines denote limits of fluctuations [197].

calm conditions prevailed below a height of 20 cm. Inoue [199] asserts that an exponential relation such as equation (16.2) can be used for wind profiles in the plant cover and he offers both observational and theoretical justification.

16.5. Profiles above Plant Covers

With a sufficiently long uniform fetch, it appears that smooth profiles of wind, temperature, and mixing ratio occur above plant covers. However, there is a number of reasons why it is difficult to obtain reliable estimates of Q_H and Q_E from such profiles.

In the first place, the roughness length z_0 and the zero plane displacement d both vary with wind speed and other factors. A field of wheat yields to the wind, creating waves as in an ocean; the analogy is not quite correct because the plants are firmly rooted. However, for particular turbulence spectra and wind speeds, resonance effects may occur and the plant may sway much more than with lighter or stronger winds. Even the direction of the wind, parallel or perpendicular to the furrows, is important in some cases [200]. The determination of the two parameters should be attempted only as a classroom exercise, and the results should not be used in any estimates of Q_H or Q_E.

Second, the three diffusivities K_m, K_H, and K_E are not likely to be equal. This is because the sources (and sinks) of momentum, heat, and water vapor are often at different levels in the plant cover. In particular, the daytime heat source will vary with the elevation angle of the sun.

Finally, the energy balance equation (1.1) cannot be used as a check on the sum $(Q_H + Q_E)$ because radiation penetrates below the canopy.

The determination of heat transfer above tall vegetation through the use of vertical profiles must therefore be approached with great caution. The eddy correlation method is preferred.

16.6. The Energy Balance within a Plant Cover

Vertical fluxes above vegetation are often assumed to be constant with height. Within the plant-air layer, however, there is certain to be flux divergence. The micrometeorologist has therefore been reluctant to take measurements below the zero plane displacement, and it is only recently that plant physiologists have persuaded him to consider the problem.

The Japanese studies have been summarized by Inoue [190]. For some height within the plant community z the energy balance of the volume below z is given by (assuming negligible heat storage)

$$(16.4) \qquad Q_N = Q_H + Q_E + Q_G$$

The quantities Q_N, Q_H, and Q_E are to be measured at height z at the top of the volume. The value of Q_N is given by the empirical equation (16.3), and there remains only the problem of separating Q_H and Q_E. Using equations (9.4) and (10.1) and assuming as a first approximation that $K_H = K_E$, it is possible to solve equation (16.4) for the diffusivity at different levels. The variation of Q_E with height may then be determined. The resulting water vapor flux within a paddy field is given in Fig. 50 [196]. The strong daytime vertical flux divergence is to be noted. In this type of analysis, the important quantity is the net radiation which provides the energy for heat transfers. However, the results are uncertain because of the assumption of equal diffusivities.

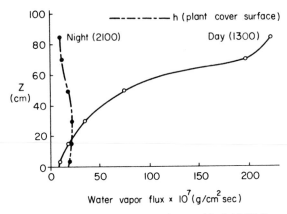

FIG. 50. Water vapor flux within a paddy field [196].

Alternatively, the problem can be viewed from the standpoint of plant physiology. Penman and Long [198] consider the flux divergence of Q_E between levels z and $(z + dz)$ as arising from the transpiration from the leaves in that layer. The upward flux of water vapor at any height z within the crop is

$$(16.5) \qquad E(z) = \int_{h/4}^{z} \Delta E$$

where h is the height of the crop; it was assumed that there were no leaves below $z = h/4$. The quantity ΔE is the horizontal vapor flux from leaves to air within the layer per horizontal square centimeter. It is determined from laboratory experiments based on stomatal population, vapor pressure differences inside and outside the leaf, etc. Some results for wheat are given in Fig. 51 for a wet soil surface (June 12) and a dry soil surface (June 16). The results are qualitatively similar to those of Fig. 50. However, two additional features are of interest. The water vapor fluxes estimated from wind profiles above the crop (dashed lines) are lower in both cases than those given by equation (16.5). In addition, on the moist-soil day, June 12, about half of the water vapor flux came from the soil.

Inoue recommends that the Japanese and the British methods

should be applied simultaneously to field observations. He also believes that the turbulent fluctuations within the crop should be studied; some preliminary work along these lines has been started

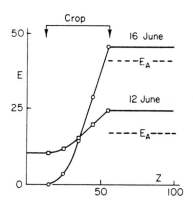

FIG. 51. Estimated upward vapor flux (ly/4 hr) on a wet soil surface day (June 12) and a dry day (June 16). Estimates of evaporation from profiles above the plant cover are given by dashed lines [198].

by the Cornell University group using hot-wire anemometers in a corn field (e.g., Wright [201, 202]).

16.7. An Alternative Notation for Fluxes

For many years the plant physiologists have used a somewhat different notation for water vapor flux, intended to emphasize an analogy with Ohm's law. Since the terminology is beginning to appear in the meteorological journals, a brief description will be useful.

The flow of water through the soil-root-plant-air system can be subdivided into a number of stages, each characterized by a rate of flow, a potential gradient, and a resistance. Then,

$$(16.6) \qquad \text{Flux} = \frac{\text{potential difference}}{\text{resistance}}$$

For example, in the turbulent boundary layer,

$$(16.7) \qquad E_T = \rho(q_1 - q_2)/r_T$$

where E_T is in gm cm^{-2} sec^{-1}, $\rho(q_1 - q_2)$ is in gm cm^{-3}, and r_T is in sec cm^{-1}. Similarly, there will be another resistance r_L in the laminar sublayer. In the case of transpiration from leaves, a stomatal resistance r_s must be added. Then the total resistance for all stages of the soil-plant-air system under steady state isothermal conditions is given by

(16.8) $$r = r_T + r_L + r_S \ldots$$

Water losses from a vegetative cover proceed both by evaporation from the soil and by transpiration from the leaves. Using the electrical analogy again, the resistances to upward moisture flux through the soil and through plants are in parallel, rather than in series ($1/r = 1/r_1 + 1/r_2$). Similarly, it may be speculated that resistances in the turbulent boundary layer and in the laminar sublayer are in series for heat or for mass transfer (actual mixing required) but are in parallel for momentum (combined effect of mixing and pressure fluctuations).

The inverse of resistance has the dimensions of a velocity. Hence, equation (16.7) is sometimes written

(16.9) $$E_T = \rho V(q_1 - q_2)$$

where V is the *transfer velocity*. Equations (10.1) and (16.9) are identical provided that

(16.10) $$V = 1/r = K/D$$

where K is diffusivity and D is called a *resistance length*. For transfer through the laminar sublayer, K is replaced by the molecular diffusivity of water vapor.

The two notations are formally equivalent, and there is some merit in the concept of an upward transfer velocity and in the consideration of the soil-plant-air system as an integrated unit. The transfer velocity is a more meaningful term to the layman than is the word "diffusivity."

Barry and Chamberlain [203] have released radioactive tracers in a wind tunnel and have measured the concentration C_z at height z, as well as the rate of uptake by plant leaves. Assuming that the rate of uptake per unit area is equal to the vertical flux F:

(16.11) $$V = \frac{\text{Rate of uptake per unit area}}{\text{Vertical difference in concentrations}} = \frac{F}{C_z - C_0}$$

The quantity C_0 cannot be measured and it is important to choose a tracer that is irreversibly absorbed by the surface ($C_0 \simeq 0$). Barry and Chamberlain infer that the transfer velocity is controlled by the rate of diffusion of the gas through the stomata. Barry [204] has subsequently released over a snow surface into the atmosphere concurrently two different tracer gases, tritiated water and radioiodine vapors. He has some evidence suggesting that the transfer velocities are in the ratio of the molecular diffusivities, about 3:1. This result is to be expected but does not affect the validity of equations similar to (10.1) in the turbulent boundary layer. If a resistance to evaporation is taking place in the laminar sublayer or because stomata are not open, the reduction in evaporation will be reflected in a comparable reduction in the vertical gradient.

16.8. Carbon Dioxide Profiles and Fluxes

Some typical carbon dioxide profiles for a field of beans are given in Fig. 52 [205]. The diurnal cycle is clearly illustrated; the upper

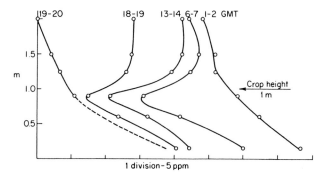

FIG. 52. Carbon dioxide profiles in and above beans, June 21, 1961 [205].

leaves were acting as a sink during the day and as a source at night; the lower leaves were either being supplied with CO_2 from the soil or (less likely) were not receiving sufficient light during the day for photosynthesis to take place.

Assuming equivalence of diffusivities, the diurnal variation of the flux of CO_2 was estimated. The results are given in Fig. 53 together with solar radiation and *photosynthetic efficiency*, the ratio of the stored energy of photosynthesis to the total solar energy Q_T. The estimated balance for the day is given in Table XVIII. Although differing in detail, a rather similar analysis has been made by Wright and Lemon [201] for a field of corn.

Carbon dioxide fluxes are important in plant physiology studies.

TABLE XVIII. Carbon dioxide flux components
for beans, June 21, 1961, mg/cm^2 day [205].

Gross photosynthesis	3.8
Respiration: Tops	0.7
Roots	0.3
Net photosynthesis	2.8
Net uptake from atmosphere	2.3
Uptake from respiration by soil	
microorganisms	0.3
Total net uptake	2.6
Net uptake estimated from mean	
dry matter production (June 16–23)	2.1 ± 0.4

They can also be very useful tracers in micrometeorology. The CO_2 gradients are relatively large and at night the source strength is almost

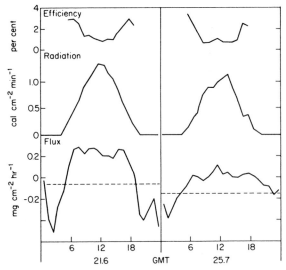

FIG. 53. Diurnal variation of CO_2 flux over beans; included also are the solar radiation and photosynthetic efficiency [205].

constant. A new method is therefore available for testing the similarity theories of micrometeorology. In particular, fast-response infrared CO_2 analyzers are being developed that will permit application of the eddy-correlation technique.

17. Forest Meteorology

17.1. The Forest: An Active Meteorological Region

Although there have been many descriptive studies of forest climatology, it is only recently that serious attempts have been made to define the exchange processes and energy balance within the stand. A standard reference is Geiger [13], while another excellent book has been published recently: "Forest Influences" [206].

Two important facts are emerging from recent studies of wooded areas. In the first place, the forest is an active meteorological region. The exchange processes are often remarkably vigorous despite the fact that winds are light. The canopy must not be considered as an impenetrable barrier for the transfer of heat and water vapor.

Second, forest climates are not all alike. Transpiration rates and soil moisture conditions differ, as well as physical characteristics such as the extent of the foliage and of the undercover. In hilly country there are also differences in slope and aspect (to be considered in Chapter 20). In short, the forest contains strong sources and sinks of heat and moisture, which depend to a large extent on physiological factors.

17.2. The Radiation Balance of a Forest

The energy available to a forest is the net radiation above the canopy Q_N, a fundamental quantity that should be measured in any field investigation. Because of the cost of erecting a tower, net radiation is sometimes measured in an adjacent clearing near ground level but differences in albedo and in $Q_{L\uparrow}$ over the two surfaces may be significant, particularly in winter with a snow cover.

A number of comparisons have been made between radiation above the canopy and that received on the forest floor. Some typical results are given in Table XIX for an oak forest near Potsdam, Germany [207].

155

When skies were overcast, the amount of diffuse short-wave radiation reaching the ground increased. The percentages, of course,

TABLE XIX. Solar radiation received on a horizontal surface at the 1-meter level in an oak forest as a percentage of incident radiation above the canopy [207].

	Clear sky	Overcast sky
Foliaged, %	9	11
Defoliaged, %	27	56

vary with forest density and may be as low as 0.1 % in the tropics [206].

Figure 54 illustrates the reduction in net radiation for a deciduous forest near Moscow in March prior to snow-melt [208]. At noon the

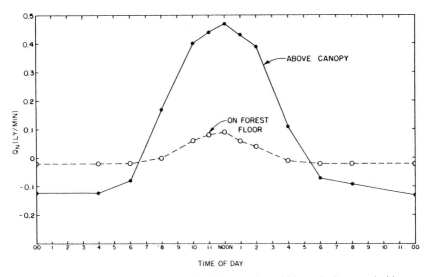

FIG. 54. Net radiation balance during clear weather within and above a deciduous forest in late winter [208].

net radiation at the forest floor was 19% of that above the canopy in clear weather.

Baumgartner [209] measured net radiation at several levels in and over a young pine forest 30 km southeast of Munich, Germany. The average diurnal cycle for 9 summer days is given in Fig. 55 while the daily totals for typical clear summer and fall days are shown in Table XX.

TABLE XX. Daily totals of net radiation in and above a young pine forest (ly/day) [209].

Height, *meters*	10.0	5.0	4.1	3.3	2.1	0.2
July 7, 1952	566	555	223	36	—	35
Percentage	100	98	39	6	—	6
November 9, 1951	291	—	104	—	14	—
Percentage	100	—	35	—	5	—

The data do not follow a Beer's law relation because of the uneven distribution of leaves with height. However, observed values of Q_N at different levels could be used in an energy balance approach similar

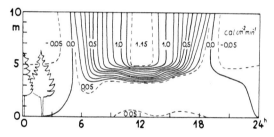

FIG. 55. Isopleths of net radiation in a young pine forest in Germany [209].

to that suggested by Inoue [190] in connection with equation (16.4); the method has recently been applied by Denmead [210] in a study of a pine forest in Australia.

17.3. Soil Temperature and Moisture

Soil temperatures depend upon whether direct sunlight penetrates to the forest floor. If it does, there can be wide variations in the upper few centimeters in both space and time. Waggoner *et al.* [211] found a difference of 9°C at the top of the forest litter within a period of 20 min. In dense stands, however, the diurnal and seasonal temperature waves are greatly damped in comparison with those in cleared areas. As a result, values of Q_G and frost penetration depths are reduced considerably in the woods.

Soil moisture is important in hydrological and forest growth studies. When a forest is cleared, what is the effect on moisture storage and run-off? A general but not invariable rule is that the upper layers are moister but the lower layers are drier in wooded country than in the open [212]. This is because of greater penetration of tree roots as compared to those of grass, with resulting greater transpiration. Although water storage capability is therefore sometimes less in a forest, wooded areas have a number of beneficial effects in terms of water management and flood control, namely:

1. Snow-cover melts slowly on the shaded forest floor in spring.
2. During heavy rains, some of the water is intercepted by the canopy.
3. Soil percolation rates are higher in a forest, thus reducing surface run-off and erosion.

17.4. Forest Temperatures

Although the forest usually has a moderating influence on temperature, exceptions do occur. In the Mediterranean region, for example, some forests actually have higher average summer temperatures than are found in open country [206]. This is explained by the fact that one particular species, *forteto*, does not transpire appreciably during the hot summer. Transpirational cooling is therefore negligible. In temperate latitudes, the beech has a higher transpiration rate than a Scots pine; a beech forest is therefore cooler than a comparable stand of pine [206].

The canopy of a well-developed forest is an active surface for transfer processes. As a result temperature profiles above the canopy show inversions at night and lapse conditions during the day. However, two additional factors must be considered. In the first place, since air can "leak" down and up through the forest, large temperature gradients (either positive or negative) cannot be maintained above the canopy. In addition, strong transpiration reduces the daytime lapse rate and in some situations may result in a temperature inversion. It has been recognized for many years by glider pilots that updrafts (free convection) are rare over a forest; the *forteto* would of course be an exception.

Temperature profiles within the forest are similar to those within a low vegetative cover, inversions during the day and lapse conditions at night. These are average conditions, and exceptions occur, particularly in an open stand or when dew forms. Some mean temperatures have been given [213] for a jungle area in Colombia. Two 200-ft towers were erected 1100 meters apart, and temperatures were measured at eight levels (Table XXI).

TABLE XXI. Mean temperatures (°C) at the north and south towers by level and time of day [213].

Time	0700–1200 LST		1200–1700 LST		1700–2200 LST		2200–0300 LST	
Level	N	S	N	S	N	S	N	S
a 6.5′	25.8	25.8	28.0	27.9	25.5	25.8	24.5	25.1
b 30′	26.0	25.9	28.4	28.1	25.6	25.7	24.5	24.9
c 56′	26.3	26.1	28.6	28.3	25.7	25.8	24.6	24.9
d 74′	26.5	26.3	28.8	28.5	25.8	25.8	24.5	24.9
e 100′	26.8	26.4	29.1	28.6	25.8	25.9	24.5	24.9
f 146′	26.8	26.5	29.0	28.7	25.9	25.9	24.6	24.9
g 170′	27.0	26.6	29.4	28.8	26.0	26.0	24.6	24.9
h 192′	26.9	26.6	29.3	28.9	26.0	26.0	24.7	25.0

The forest was most dense just above the *e*-level; it was quite open at the *f*-level; the *g*- and *h*-levels were completely above the canopy. The temperature gradients show the expected trends. An additional result of great interest is that mean temperatures at the two towers were quite similar. Baynton applied a Student's *t*-test and concluded that a single tower yielded representative mean temperatures at his site.

Heckert [207] has studied the effect of variable cloudiness on temperature fluctuations at 8 levels on a tower in an oak forest near Potsdam, Germany. The mean amplitudes of more than 2000 cases of short duration ($t < 15$ min) are given in Table XXII.

In summer the temperature wave originates at tree crown level (14–15 meters), and propagates both up and down with decreasing

TABLE XXII. Mean amplitudes of temperature waves in °C at 8 levels in foliaged and defoliated forest (average tree height of 15 meters) [207].

Height, meters	0.5	2	10	12	14	15	20	25
Foliaged forest	2.0	2.0	2.2	2.3	2.6	2.5	2.2	1.9
Defoliated forest	1.6	1.2	1.1	1.1	1.2	1.2	1.0	0.9

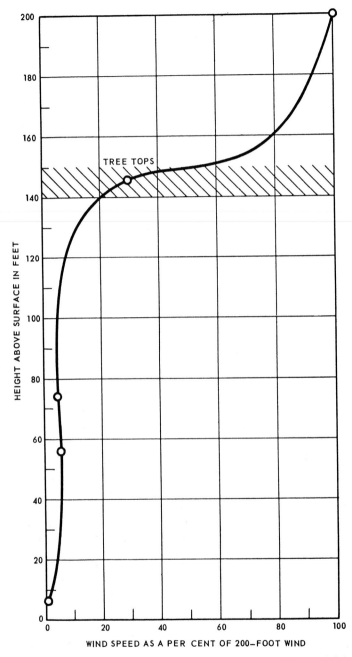

Fig. 56. Mean wind speed as a percentage of 200-ft wind speed versus height in the Colombia jungle [213].

amplitude. The fact that it can be detected on the forest floor is strong evidence for an active heat exchange process within the stand. When there are no leaves on the trees, both the crowns and the forest floor are important interfaces for the initiation of temperature waves. In this case the amplitudes are at a maximum at 0.5 and 14–15 meters.

Bayton [213] has found a similar effect in the Colombia jungle. He concludes that there is a large vertical flux of heat within the forest.

17.5. Winds in the Forest

Winds above the canopy can be approximated by a logarithmic or power law form. Rauner [208] has found a considerable variation in roughness length z_0 and zero-plane displacement d in a deciduous forest of mean height 17 meters. Some results are given in Table XXIII. As wind speed increases, the zero-plane displacement decreases but the roughness length increases.

TABLE XXIII. Values of z_0 and d for 3 wind speed classes above the canopy in winter [208].

Wind speed, meters/sec	2.0–3.5	3.5–5.0	5.0–7.0
z_0, meters	2.7	3.2	5.0
d, meters	8.5	6.0	4.0

Wind flows below the canopy are light and disorganized. Bayton [213], using two towers 1100 meters apart in the Colombia jungle, found no correlation in wind direction below the top of the canopy. However, at the 146- and 200-ft levels (at and above the tree tops), the correlation coefficients were statistically significant. Average wind speed profiles are given in Fig. 56, which are qualitatively in agreement with results from temperate zones.

If exchange processes are active within the forest despite the fact that winds are light, the turbulent fluctuations must be of considerable importance. Unfortunately, fast-response data are not yet available.

17.6. Humidity in the Forest

When the soil is dry, the principal water vapor source is the tree crown. Mixing ratios then decrease upward and downward from that level during the day. When transpiration ceases at night, the profiles tend to readjust to a condition in which the mixing ratio decreases upward from the ground.

When the soil is moist, a second active surface for evaporation is present. In addition, there may be undergrowth that is transpiring. The resulting daytime mixing ratio profiles within the forest can therefore be quite variable in time and space.

17.7. Heat Storage within Trees

Tree trunks, branches, and leaves are a small but significant fraction of a forest, particularly in the tropics. Baumgartner [209] estimated that the biomass of a pine forest near Munich was equivalent to a layer of wood over the entire surface to a depth of 1.89 cm. The heat capacity of living wood is about 0.5 to 0.65 cal/deg cm^3, and the possibility of the biomass contributing to the heat storage of the forest must not be overlooked.

Baumgartner made a few measurements of tree trunk surface temperature and found values almost identical with adjacent air temperatures. In the crowns, however, tree surface temperatures were 8°C higher by day and 0.2°C cooler by night than surrounding air temperatures. Baumgartner assumed that the higher skin temperatures were compensated by lower interior temperatures within the tree; he then computed heat storage from air temperature profiles within the stand, using an equation similar to (15.4).

$$(17.1) \qquad\qquad Q_S = \rho c h_1 \int_0^h \frac{\partial T}{\partial t}\, dz$$

where ρc is the heat capacity of living wood, h_1 is the equivalent thickness of a layer of wood equal to the forest biomass, and h is treetop level.

Although the air within the stand occupies a much larger volume than does the biomass, its heat storage is a much smaller quantity. A numerical value can be calculated from equation (17.1) using the appropriate heat capacity for air and deleting h_1. For the pine forest in Germany, Baumgartner obtained the results given in Fig. 57.

The possibility has been suggested recently of estimating transpiration rates from sap speed in the trunks [214]. It has been demonstrated, for example, that within 24 hr after dry soil is irrigated, the flow rate of sap in juniper trees increases considerably. Within the next few years, it may be possible to obtain an independent estimate of transpiration rates by this method.

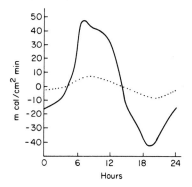

FIG. 57. Heat storage rates of the forest biomass (solid line) and of the air within the stand (dotted line) [209].

17.8. The Energy Balance of a Forest

Baumgartner [209] and Rauner [208, 215] have calculated the energy balance of a forest volume.

$$(17.2) \qquad Q_N = Q_G + Q_H + Q_E + Q_S$$

where Q_N, Q_H, and Q_E are fluxes above the canopy, Q_G is the soil heat flux, and Q_S is the forest heat storage.

Because of instrumental difficulties in determining moisture

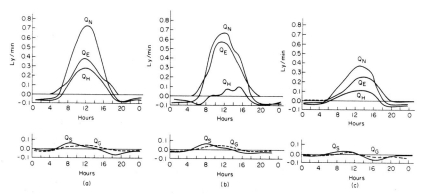

FIG. 58. The energy balance of a deciduous forest in summer in the U.S.S.R. (a) clear dry day, (b) clear day after rainfall, (c) cloudy day [215].

profiles, Q_E was assumed to be the unknown term by Rauner in his study of a deciduous forest 85 km north of Moscow in late summer. Some results are given in Fig. 58 for a clear dry day, a clear day after rainfall, and a cloudy day. The transpiration rates in the three cases were estimated to be 2.4, 4.0, and 2.0 mm/day, respectively. The difference between a clear dry day and a clear day after rain, both having similar values of Q_N, is through changes in Q_H and Q_E; the soil heat flux and the heat storage terms are not greatly affected.

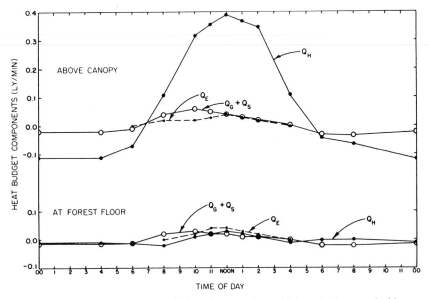

FIG. 59. The energy balance during clear weather within and above a deciduous forest in late winter (see Fig. 54 for net radiation component) [208].

Rauner also computed the heat balance of a snow-covered forest in late winter. He assumed that transpiration was negligible but that there was evaporation from the snow surface. Vapor pressure and net radiation measurements were made at the forest floor, and Rauner was able to estimate the energy balance above and below the canopy. The results are given in Fig. 59, which shows the partition of available energy Q_N (from Fig. 54) among Q_H, Q_E, Q_G, and Q_S. Above the canopy, the eddy heat flux Q_H was the dominant term; on the forest floor, all components were small. Assuming equivalence of diffusivities of water vapor and heat, it was possible to estimate Q_E from vapor pressure differences as well as from equation (17.2) as a remainder term. The two estimates agreed quite well.

Baumgartner [209] used wind and temperature profiles above a pine forest to calculate Q_H. The quantity Q_E was then obtained as the remainder term in equation (17.2). The average values for a summer period are given in Fig. 60; the partition of energy is quite similar to that given in Fig. 58.

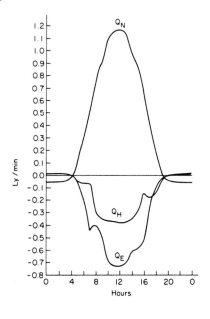

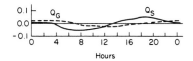

FIG. 60. The energy balance of a young pine forest in Germany (average for June 29-July 7, 1952); the quantities Q_H and Q_E are opposite in sign by author's convention [209].

Denmead [210] has estimated flux divergences on two occasions in a pine forest in Australia (mean tree height was 5.5 meters), while Philip [216] has used Denmead's data to test a model for forest heat transfer. In both papers, K_H and K_E are assumed to be equal. Denmead finds that transpiration rates at different levels are roughly proportional to foliage density; in the lower layers (1.5 to 3 meters), the transpiration rate is in excess of available radiant energy, the difference being supplied by a flux of sensible heat from above.

17.9. Additional Remarks

The microclimates of forest clearings and of shelter belts will be considered in Chapter 22. Within the forest itself, there is evidence of an active vertical exchange mechanism despite the absence of an organized wind flow. It is therefore imperative that measurements be made of the turbulent fluctuations of wind, temperature, and vapor pressure above and below the canopy. The necessary fast-response instruments are becoming available, and every effort should be made to obtain estimates of the variances and covariances of turbulent energy.

Despite the requirement for fast-response data, useful estimates of forest heat balance can be obtained with presently available equipment. The most important parameter is net radiation (and its vertical flux divergence). This information, in conjunction with vertical profiles of temperature and of mixing ratio, is recommended for energy balance studies.

There has been a reluctance by some to equip a single tower with expensive instruments because it is not known whether measurements at a point are representative of conditions even a few meters away. Carefully designed experiments may reveal that sensors should be on moving tracks to yield space averages at different elevations.

18. The Air over Oceans and Large Lakes

18.1. Introduction

Air and water are turbulent media obeying the same general laws of fluid mechanics. For example, the problem of diffusion of stack gases is analogous to that of the dilution of liquid wastes released from a submerged pipe. The mathematical formulations are identical.

The oceanographer, the limnologist, and the meteorologist all study the motion of fluids. They find themselves with a common problem at the air-water interface. It is well known that strong winds produce ocean waves; it is not always realized that waves affect the structure of the atmospheric boundary layer. In particular, the spectra of waves and of air turbulence are interrelated.

Important interactions take place at an air-water interface, and there are doubts as to whether the theory developed for flow over unyielding solid ground still applies. The energy balance, too, must have distinctive characteristics because of such factors as the heat capacity of water, the presence of turbulence in the sea, the penetration of solar radiation below the interface, and evaporation of spray.[4]

18.2. Some Physical Properties of Oceans and Lakes

Water is almost incompressible, the adiabatic lapse rate being less than 0.2°C/1000 meters as compared with 1°C/100 meters for the atmosphere. The density of water depends on temperature and salinity, the maximum density occurring for pure water at a temperature of 3.94°C. As salinity increases, both the freezing point and the temperature of maximum density decrease slightly. The saturated

[4] An excellent book on maritime meteorology has recently (1965) been published by Academic Press: "Physics of the Marine Atmosphere" by H. U. Roll.

vapor pressure over sea water is about 98 % of that over pure water at the same temperature.

Consider the consequences of heating a lake at an initial uniform temperature of 1°C. The warmed surface water becomes heavier and sinks, mixing the lake thoroughly until temperatures at all levels reach 4°C. Thereafter, heated surface water is lighter than water at deeper levels, and further vertical motions are repressed. Heat transfer continues by the turbulent mixing action of wind-generated waves (in the top few meters) and slowly by conduction (throughout the fluid). In summer, therefore, a body of water tends to be divided into an upper relatively warm layer, called the *epilimnion*, and a deep, cold, and undisturbed region called the *hypolimnion*. The plane of separation is the *thermocline*, defined as the level of the maximum rate of decrease in temperature. The hypolimnion may be absent in shallow lakes.

The depth of the thermocline shows considerable variation through-out the day and the season, due to changes in the surface energy balance, the strength of the wind-generated waves, and tides. In the case of a closed basin, there will also be *seiches*, resonant oscillations alternately piling up water on the windward and leeward ends of the basin. The period of oscillation ranges from minutes to hours de-pending upon the dimensions of the lake. The variation in thermocline level is usually greater than changes in surface water level.

In autumn when surface water cools to a temperature of 4°C, its density increases and it sinks. Cooling of the entire epilimnion pro-ceeds rapidly. However, additional cooling below the temperature of maximum density brings a return of stable stratification, and subse-quent heat losses to the air are reduced. Meanwhile, conductive heat transfer is still warming the hypolimnion; in very deep lakes the tem-perature near the bottom may not reach its maximum value until midwinter.

The freezing of open water is difficult to predict. The factors of importance are salinity, wind speed, and the initial temperature structure of both air and water. A deep lake is a large reservoir of heat; wave action will therefore promote vertical heat exchange in the water and inhibit the formation of ice. On the other hand, when the wind is light and the water surface smooth, the air is quickly modified, reducing the air-water temperature difference. There is a large difference in heat capacities of air and water. In the absence of evaporation, Ovey [217] has calculated that cooling of a meter-deep layer of water by 0.1°C would raise the temperature of a 30-meter thick layer of air by 10°C.

Two other physical properties of lakes and oceans should be mentioned:

1. Mean flows (water currents) are weak except in rivers. The turbulent components, on the other hand, are often large.
2. Bodies of water contain horizontal temperature gradients, which have obvious meteorological significance. It is only when wind flow is parallel to the surface water isotherms that advection effects can be ignored. Air blowing across an open prairie is not modified greatly but air blowing at right angles to the sea isotherms in the North Atlantic Ocean will be continually heated or cooled from below.

Horizontal temperature gradients are particularly large near shore-lines during periods of *upwelling*; if winds are blowing from land to sea in summer, the warm surface water is carried offshore to be re-placed by cool water from the ocean bottom.

18.3. The Radiation Balance of Oceans and Lakes

Long-wave radiation is absorbed by a very thin film of water. In contrast with a solid surface, however, short-wave radiation pene-trates to some considerable depth in a lake or ocean. Similarly, albedo values must include diffuse solar energy that is being reflected upward from a level of perhaps 5 or 10 meters below the surface.

Short-wave radiation penetration creates a problem similar to that found in connection with snow and ice surfaces. The application of equation (1.1) leads to a substantial "remainder" term and it is necessary to consider the heat budget of a volume. Additional com-plications arise because the term Q_G must include turbulent transfer of heat by the water, and because there may be horizontal advection of heat out of the volume by water currents.

When skies are clear, the albedo of a water surface is 2–3% for solar altitudes greater than 45° increasing to 40% at 5°. Under cloudy skies, the ocean has an average albedo of about 8–10% because diffuse solar radiation is arriving at the sea surface at many different angles of reflection. The roughness of the water does not appear to have any great effect on albedo, although this is not firmly established.

The attenuation of solar radiation by water can be approximated by Beer's law, equation (2.3). In clear tropical water some energy penetrates 700–1000 meters but in general the amount contributing significantly to local heating is negligible below 10 meters. There is a dependence upon wavelength, which explains the variation in color of different bodies of water. Figure 61 displays the percentage of

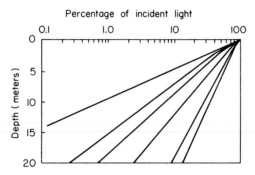

Fɪɢ. 61. Spectral distribution of light in Lake Huron for wave bands from top to bottom of diagram of 0.61–0.75, 0.30–0.43, 0.59–0.61, 0.54–0.59, 0.49–0.54, and 0.44–0.49 μ, respectively [218].

incident light in different wavelength bands reaching different depths in Lake Huron [218]. Extinction coefficients are given in Fig. 62 for all of the Great Lakes as well as for a small inland lake. A large value of the extinction coefficient (Lake Erie and Frains Lake) corresponds to strong absorption of solar radiation near the surface.

Net radiation is an important component of the energy balance of a lake or sea. Unfortunately, it is very difficult to obtain experimental measurements. In the first place, the sensor must " see " only the water surface, not the side of a ship or buoy. Second, the radiometer must be kept level. Finally, spray must not be allowed to reach the instrument. The net radiation of small lakes can be measured successfully; however, the only feasible method for the open sea appears to be aircraft observations. Most energy balance studies of oceans use empirical estimates of net radiation.

18.4. Temperature and Humidity near the Water Surface

Since evaporational cooling can create large temperature gradients in the upper few millimeters of water, it is necessary to distinguish between *surface temperature* (measured with a bucket or with a temperature sensor mounted on a buoy) and *surface radiative temperature* (measured with a radiometer).

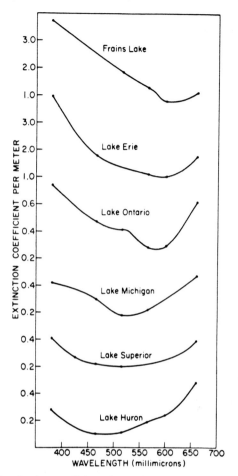

FIG. 62. Vertical extinction coefficients for various wavelengths in the Great Lakes and Frains Lake (a small inland lake) [218].

The diurnal variation of temperature is much less in water than in soil because of turbulent wave action. Figure 63 displays mean values of temperatures at the surface and at a depth of 10 meters [219]. The amplitudes are 0.2 and 0.15°C, respectively, and there is a slight lag with depth. In the absence of advection, therefore, there is no mechanism to create atmospheric superadiabatic or inversion conditions, and lapse rates remain close to the adiabatic, although frequently showing minor irregularities [220].

When warm air blows across cool water, as so often happens over the Grand Banks of Newfoundland in summer, an intense advection inversion forms. On the other hand, when cold air blows across warmer water, free convection develops and there is a large vertical heat transfer rate; the superadiabatic lapse rate is usually restricted to the lowest 10 meters [221].

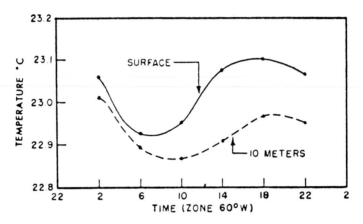

Fig. 63. Mean diurnal variation in water temperature, June 12–23, 1953 at 33° 33′N, 62° 26′W [219].

These results refer to data obtained from buoys and give no indication of what is happening very close to the interface. Hamilton [222] measured temperature profiles in and over Lake Mendota, Wisconsin using a fast-response thermocouple on a vertical lift moving at a rate of 14 cm/sec. The sweeps extended from 200 cm above the surface to 80 cm below. It was possible to obtain meaningful values across the interface on descent but not during an upward sweep because of evaporational cooling of the sensor. The fetch was 2–3 km.

Most of the profiles showed a sharp discontinuity at the air-water interface, whether the air was warmer than the water or vice versa. An example is given in Fig. 64 (oscilloscope photograph) for a day when the water temperature at a depth of 15 cm was 3.4°C colder than the air at a height of 20 cm. Gates *et al.* [223] have measured the vapor pressure gradient over a free water surface in a laboratory in the absence of wind, using a microwave refractometer. There was a large vapor pressure gradient in the lowest 10 mm; in the 10 to 30 mm layer, on the other hand, the gradient was small.

In order to obtain even finer resolution of the interface temperature structure, it is necessary to use radiometers. McAlister [224] measured radiative temperatures in two wavelengths at the Scripps pier in California. He found that the top 0.02 mm of water was 0.2°C lower than that of the top 1 mm.

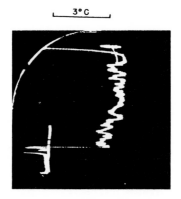

FIG. 64. A single downsweep of a fast-response thermocouple into Lake Mendota, Wisconsin from a height of 200 cm (top of picture) to water surface (discontinuity) to a depth of 80 cm (bottom of picture) [222].

Many estimates of Q_H have been made using shipboard-air/sea-surface temperature differences. In addition, shipboard vapor pressure and saturated vapor pressure at the sea-surface temperature are used to determine Q_E. The method has been necessary because of the urgent need for estimates of heat flux and evaporation in ocean areas where there are no vertical profiles of wind, temperature, and humidity. There is, of course, no independent way of verifying the results. Since surface radiative and sea-surface temperatures have been shown to be different, there is a need for some carefully planned experiments relating temperature gradients from buoys with shipboard-air/sea-surface-radiative temperatures. It may be that the sea-surface/sea-surface-radiative temperature difference will prove to be an index of the rate of evaporation.

Specific humidity usually decreases upward from the water. There are not yet enough data to establish similarity of humidity and temperature profiles. During strong winds, spray may contribute appreciably to the water vapor content of the air, in which case K_E could be quite different from the other diffusivities.

Hasse [225] has examined profiles of wind, temperature, and humidity measured from a buoy in the Baltic and North Seas for several levels between 0.8 and 13.6 meters. For 150 cases in which the lapse rate was adiabatic, the potential temperature was assumed to be the surface radiative temperature T_0. This value proved to average 0.1 to 0.3°C lower than the sea temperature T_w at a depth of 0.5 meter, the difference increasing with increasing Q_E, as computed from equation (10.2).

Other cases were examined for which the humidity gradient was negligibly small. Assuming that the vapor pressure was equal to the saturated vapor pressure at the surface, it was possible to infer a surface temperature T_0. In superadiabatic cases, the water temperature T_w was higher than T_0, and the data suggest an upper limit of about 1°C for $(T_w - T_0)$. During inversions, T_w was lower than T_0 but there appeared to be no limiting maximum for the difference. This may be explained by differences in stability in the upper few centimeters of the ocean.

18.5. Wind Flow over Water

For obvious reasons, it is almost impossible to obtain meaningful wind measurements within a meter or so of a disturbed water surface. In the 1- to 16-meter layer, the wind profile is often logarithmic, although Klevantsova *et al.* [220] have recently reported that departures were significant and frequent in their data. It must also be recognized that results are biased toward cases of rising sea with few observations in a falling sea after a storm. Finally, Sheppard [74] suggests that the latent heat effect on buoyancy mentioned in Section 12.4 may be important at times over water so that an adiabatic lapse rate is not a condition of neutral stability. Sheppard assumes that any logarithmic wind profile is indicative of an atmosphere in the neutral state.

Waves and wind are rarely in equilibrium except for small lakes, and it is intuitively evident that frictional drag depends in some way on whether the waves and the wind are moving in the same or in opposite directions. Published wind profiles often do not include concurrent information about the state of the sea. As a result, estimates of the roughness length and of the drag coefficient display a wide scatter.

Values of the roughness length z_0 range from 1×10^{-4} to 50 cm over water. Most of the estimates have been summarized by Kondo [226], who suggests that for logarithmic wind profiles:

$$(18.1) \qquad z_0 = 1.4 \times 10^{-7} \bar{u}_{10}$$

where $\bar{u}_{10}$ is the mean wind at a height of 10 meters. The relation is valid only for the wind speed range of 10 to 30 meters/sec and even then is subject to large experimental scatter. As in the case of wind flow over tall vegetation, the estimation of z_0 and of d is rather meaningless except possibly for climatological purposes. Equation (7.11) is particularly sensitive to errors in the calibration of anemometers when used for estimating z_0. The slope of the wind profiles, on the other hand, will yield the friction velocity and the drag coefficient with rather more success. Deacon and Webb [227] have summarized experimental estimates of C_D and suggest an empirical equation:

$$(18.2) \qquad C_D = (1.00 + 0.07\bar{u}_{10}) \times 10^{-3}$$

Sheppard [74] has proposed a similar expression on the basis of measurements taken on Lough Neagh, Northern Ireland where the fetch was about 25 km:

$$(18.3) \qquad C_D = (0.12 + 1.2\bar{u}_{10}) \times 10^{-3}$$

Since a water surface yields to the wind it is likely that the defining equation for drag coefficient [equation (7.2)] should be modified. Sheppard speculates that the following empirical form might be useful:

$$(18.4) \qquad \tau_0 = C_D\rho(\bar{u}_{10} - \bar{u}_w)^2$$

where $\bar{u}_w$ is wave velocity. The drag is small when the sea is smooth. When the first ripples appear as the wind increases, the water motion may actually reduce the drag. With a further strengthening of the wind, of course, the drag becomes larger. When the wind is less than 3 meters/sec, the drag may also be affected by biological contamination (creating streaks or slicks) [228].

Vinogradova [229] has found a dependence of C_D upon lapse rate, smaller values occurring during inversion conditions for the same reference-level wind speed. She has also found good agreement in values of shearing stress obtained from wind profiles and by the eddy correlation technique, equation (8.23), using a hot-wire anemometer at a height of 2 meters.

The question still remains as to whether it is justified to apply theory developed for fluid flows over a rigid surface to cases where the underlying medium is in motion. The transfer of momentum at the interface is therefore a question of considerable importance, particularly when wind and waves are not in equilibrium. The theoretical models proposed by the oceanographers suggest that energy transfers are largely by resonance mechanisms, but as yet the models contain too many simplifying assumptions to yield precise details.

On theoretical grounds, it is clear that the logarithmic wind profile cannot apply below the level at which the mean wind velocity is less than the water wave phase velocity [230]; no wind measurements between wave crests have been made, but Stewart believes that air motions in this region are organized and wavelike, presumably contributing little to heat and water vapor fluxes. Since momentum is transferred by pressure fluctuations whereas heat and water vapor fluxes require actual mixing, it follows that K_m is likely to be quite different from K_H and K_E over a water surface. Sheppard [155], on the other hand, does not feel that this argument is convincing.

18.6. Heat Storage and Horizontal Advection in Water

Heat gain or loss by a column of water of depth D is given by

$$(18.5) \qquad Q_s = c \int_{-D}^{0} \rho \frac{\partial T}{\partial t} \, dz$$

The density and specific heat are usually assumed with sufficient accuracy to be equal to unity.

The quantity Q_s is called the *heat storage* and can be found from consecutive temperature profiles in the lake or ocean. Temperature changes with time depend upon the heat exchange at the upper surface, conduction, the vertical component of turbulence within the water, advection, and the penetration of short-wave radiation.

In the case of a closed basin, the cross-sectional area decreases with depth. Equation (18.5) must then be normalized to a unit column of converging sides; details have been given by Dutton and Bryson [231].

Horizontal advection by ocean currents is usually small. When necessary, however, it can be determined with sufficient accuracy by the equation:

$$(18.6) \qquad Q_v = Mc(T_1 - T_0)$$

where Q_v is horizontal advection, M is the integrated water-mass transport across the boundaries, and T_1, T_0 are the mean vertical temperatures at each boundary. In a study of the energy balance of the Caribbean Sea, Colón [232] obtained values of Q_v from equation (18.6) ranging from 28 ly/day in August to -13 ly/day in December. For Lake Ontario, Bruce and Rodgers [233] estimated that Q_v ranged from -12 ly/day in January to 2 ly/day in May.

18.7. Energy Balance Calculations of a Lake or Ocean

Equation (10.2) is commonly used for evaporation estimates over large bodies of water, assuming no advection and an adiabatic lapse rate:

$$(18.7) \qquad E = \frac{k^2 \rho (q_s - q)\bar{u}}{(\ln z/z_0)^2}$$

where q_s is the mixing ratio of saturated air at sea-surface temperature, although strictly speaking it should be for the level z_0. There are many inaccuracies introduced by assuming adiabatic conditions, equivalence of diffusivities, etc.

Swinbank [234] suggests that equation (18.7) is of the general form:

$$(18.8) \qquad E = B\bar{u}(q_s - q)$$

where B is a slowly varying parameter. In view of all the other assumptions and approximations, Swinbank suggests the use of a constant value for B of 0.24×10^{-5}, when $\bar{u}$ and q are measured at a height of 8 meters and are in cgs units.

Equation (18.8) will underestimate evaporation when there is cold air advection and will overestimate during inversions. More sophisticated models may be used but it cannot be fairly stated that quantitative values of Q_E are possible from shipboard observations.

Another method of estimating Q_H and Q_E that has not been widely exploited (largely because of cost) is the determination of vertical profiles by radiosondes or aircraft on the upwind and downwind sides concurrently of lakes or ocean strips. Changes in the heat and moisture content across the volume of air can be used to infer the vertical fluxes at the water surface. Palmén [235] has used the method to estimate evaporation from the Baltic Sea while Ninomiya [236] has undertaken a similar study of the Japan Sea.

18.8. Turbulence over Water

Eddy correlation methods will provide good estimates of heat flux and evaporation from water surfaces when the experimental difficulties are overcome. Some preliminary studies have been reported from the Soviet Union, the United States, and Germany [229, 237, 238].

Slade [239] has studied the effect of a 7-mile water trajectory across Chesapeake Bay in the United States upon wind speed and horizontal wind direction *range*, where range is defined as the difference between extreme azimuth values in degrees over a 5-min period. Some results are given in Table XXIV.

TABLE XXIV. Wind speed and direction range ratios on the upwind and downwind shores of Chesapeake Bay [239].

	No. of cases	Ratio of upwind to downwind speeds	No. of cases	Ratio of upwind to downwind ranges
Day				
Warming				
from below	49	0.69	52	1.81
Neutral	23	0.75	25	2.18
Cooling				
from below	54	0.86	54	2.34
Night				
Warming				
from below	110	0.50	118	1.79
Neutral	26	0.61	30	2.38
Cooling				
from below	24	0.83	24	2.39

The water trajectory caused the wind speed to increase, particularly during cold air advection, although there was one earlier occasion (not included in Table XXIV) when the water was more than 15°C cooler than the air and the intense inversion resulted in a wind speed of half its upwind value.

The wind direction ranges diminished with an over-water trajectory, the effect being greatest during advection inversions. These results are indicative of the stabilizing effect of water on the diffusion of a plume.

19. Land and Sea Breezes

19.1. Introduction

During periods of light geostrophic winds in spring and summer, surface winds blow from ocean to land during the day (sea breeze) and in the reverse direction at night (land breeze). In winter the land breeze may occur in the daytime as well as at night. These are, of course, generalities that require considerable modification in specific situations.

There are a number of reasons why "coastline meteorology" is of practical importance. First, many large cities are located on bodies of water, either because of their historic role as seaports or because of the need for water for industrial processes. Second, the shoreline has always been a favorite holiday area. Finally, coastal winds are of importance to shipping, particularly small craft. Marine forecasts, while accurate enough over open water, may be seriously in error just off shore, especially if the coast is rugged.

The line of separation between land and water constitutes a discontinuity in meteorological boundary conditions. The roughness of the underlying surfaces differs, as well as do heat and moisture conditions. Since the diurnal and annual cycles of wind, temperature, and humidity are much smaller over water than over land, interactions are to be expected near the coastline. Of some importance also is the upwelling of cold water during land breezes. Whereas sea-surface temperatures change slowly in the middle of the ocean, wind shifts can produce rather sudden falls or rises in water temperature near coastlines.

The properties of the transitional zone depend upon a number of factors:

1. The land-water temperature difference.
2. The strength and direction of the geostrophic wind.
3. The time factor (the transitional zone gradually widens).

4. The roughness of the terrain, including the presence of cliffs and sloping land, and the roughness of the water.
5. The curvature of the coastline, the limiting cases being a small island and a small lake.
6. Moisture conditions over land, e.g., forest versus desert.

The land-sea circulation is often considered to be a mesometeorological phenomenon. However, the sea breeze commences on a very local scale and is of considerable interest to micrometeorologists.

19.2. Land and Sea Breezes during Light Geostrophic Winds

Consider the simplest case in which the geostrophic wind is calm, skies are clear, and the shoreline is straight. The daytime development of a sea breeze has been described by Pearce [240] in what he suggests as a textbook definition.

> "When the air receives heat from the land surface, it is distributed vertically through a layer some hundreds of feet thick by convection currents. The consequent expansion which proceeds at the same time causes a tidal motion from land to sea which extends through the whole depth of the atmosphere, rather like the motion in a long water trough caused by raising one end very slowly, or as one could imagine would take place if the land surface was raised instead of the air heated. The tidal velocities are small, of order 1 km hr^{-1}. Another kind of motion also occurs at the same time; there is a horizontal density gradient across the coastline, and gravity acting on this density gradient causes the heavier colder air over the sea to flow inland underneath the warmer air, the velocity in this case being of order 10 km hr^{-1}. The circulation in the vertical plane must be continuous, and this entails a return motion of the warmer air from land to sea at higher levels, with a maximum velocity of about 5 km hr^{-1} just above the forward edge of the current of cold air. This local circulation, which is the observed sea breeze, transfers mass from sea to land."

During the day the transitional zone widens. As air trajectories lengthen, the Coriolis force of the earth's rotation becomes important, causing winds to veer until by late afternoon the direction of flow is parallel to the coast.

When the circulation commences, there is warm air over land and cool air at sea. However, the temperature structure is modified in two ways:

1. Subsiding air is warmed adiabatically. If, for example, there is a slight inversion over the sea at sunrise, the inversion will intensify at the levels where subsidence is taking place.
2. The sea air moving inland is heated from below. A new internal boundary layer develops in which the lapse rate is superadiabatic.

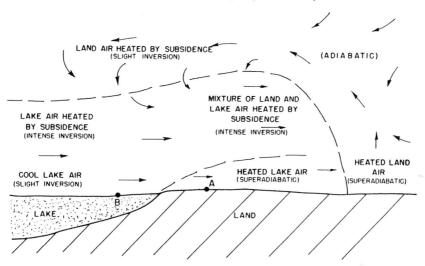

Fig. 65. Schematic representation of a sea breeze when the geostrophic wind is light.

The resulting daytime sea-breeze structure is shown schematically in Fig. 65. Typical vertical temperature profiles were displayed in Fig. 19; the inversion base was the top of the internal boundary layer, not the lake-breeze frontal surface; the latter was at a height of about 3000 ft.

In the initial sea-breeze surge an hour or so after sunrise, there is a sharp drop in temperature (2–10°C) and rise in humidity. Later in the day and at greater distances inland, the arrival of the sea breeze is indistinct because of modification by the heated ground. In fact, it may disappear completely, permitting a secondary sea-breeze front to develop and move inland [241].

The vertical dimensions of sea breezes have been investigated by glider pilots [241]. In one specific case (July 6, 1956 near Lasham, England about 50 km from the sea), there was a narrow belt (100–250 meters wide) just ahead of the sea-breeze front in which there were lifts occasionally as large as 8 meters/sec. The vertical extent of the frontal surface appeared to be about 1.5 km.

At night when land is cooler than water, a reverse circulation develops. However, it is weak because of the associated nocturnal radiation inversion. The land breeze usually displays intermittency: short bursts of wind between periods of relative calm. If the wind does blow steadily, upwelling of cold water results and the land-sea temperature

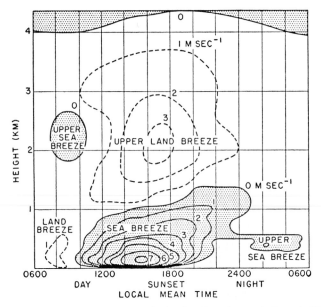

Fig. 66. Wind-speed isopleths for the land and sea breeze in Batavia [242].

difference is reduced. Air trajectories at night are rarely of sufficient length for the Coriolis force to become important. The land breeze therefore usually blows at right angles to the coast.

A widely quoted example is displayed in Fig. 66, the land- and sea-breeze regime at Batavia [242]. The daily reversal in flow is clearly illustrated.

When the shoreline is rugged, winds and temperatures are subject to many microscale variations. Dexter [243] used an automobile to obtain the temperature distribution in Halifax, Canada during sea-breeze days. An example is given in Fig. 67 for May 7, 1953, 1300–1330 AST. The temperature varied from 56° to 70°F throughout the city.

Slope and valley winds (Chapter 20) usually occur in conjunction with land-sea breezes to complicate the flow patterns. This is one reason why there is such a variation in the reported horizontal width of the transitional zone. The other reason, of course, is the modification of sea air as it moves inland so that its arrival often passes without notice.

Sea breezes are more intense when the land is a desert than when it

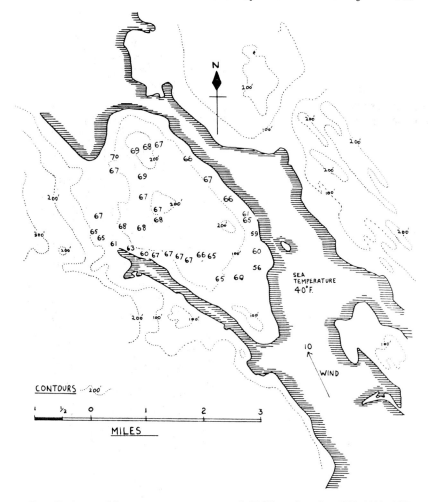

Fig. 67. Automobile temperature traverse of Halifax, Canada, 1300–1330 **AST**, May 7, 1953 when a sea breeze was blowing [243].

is forested. A wide sandy beach may at times become sufficiently hot to create a double circulation cell, with air flowing over the beach from both land and sea.

19.3. Sea Breezes When a Geostrophic Wind Is Blowing

When the geostrophic wind is at right angles to the coastline and from water to land, no sea breeze is possible in the usual sense of the word.

As described in Chapter 12, a new boundary layer begins to develop at the shoreline and thickens inland. It is theoretically possible for a land breeze to develop at night and overcome the geostrophic wind; however, this rarely happens except in conjunction with slope or valley winds.

When the geostrophic wind is blowing from land to water, the nighttime land breeze is intensified, and a new boundary layer develops offshore. The daytime sea breeze may or may not develop, depending upon the strength of the geostrophic wind, the amount of differential heating, and other factors. If it does form, its arrival is usually accompanied by a sharp wind shift, drop in temperature, and rise in humidity. For the first few hours after sunrise, the heated land air moves in the direction of the geostrophic wind, pushing back the cool sea air. If a sufficient pressure differential develops, the sea breeze moves toward the shore, undercutting the land air, and developing the characteristics of a macroscale cold front. A number of attempts have been made to develop criteria for predicting sea breezes with off-land regional winds and to forecast the inland limit of their penetration. However, no general rules can be given.

19.4. A Mathematical Model of the Sea Breeze

A number of solutions have been presented for the equations of motion for differential heating of land and sea. Because of the complexity of the problem, simplifying assumptions must always be introduced. In particular, the sea-breeze circulation extends up above the layer of constant shearing stress. Nevertheless, considerable progress has been made in recent years through the use of high-speed electronic computers, which perform rapid numerical integrations and which make it possible to reduce the number of simplifying assumptions.

One of the more realistic models appears to be that of Estoque [244, 245]. It must be recognized, however, that although Estoque's assumptions and solutions are reasonable, they cannot yet be compared with observations. The only wind and temperature measurements that are sufficiently complete in time and space are by Fisher [246] and Frizzola and Fisher [247] but the macroscale situation was rather unsteady, the coastline was irregular, and an island disturbed the temperature field.

Estoque considered a sea-breeze circulation bounded by a vertical cross section 2 km in depth and extending horizontally ± 200 km from the coastline in both directions. Below a height of 50 meters, the fluxes of momentum and heat were assumed to be constant, and the usual boundary layer theory applied. Above the 50-meter level, diffusivity was assumed to decrease linearly to a value of zero at a height of 2 km. The equations of motions in this upper layer included horizontal pressure gradient and Coriolis force terms. Boundary conditions were prescribed by a constant temperature over the water but by a sine-wave land temperature function.

Finite difference integrations were performed at consecutive time increments of 5 min, commencing at 0800 LST. A number of solutions were obtained for different geostrophic wind directions and initial lapse rates. Some typical results for 1700 LST are given in Figs. 68 to 70. The numerical prediction in Fig. 68 (zero geostrophic wind)

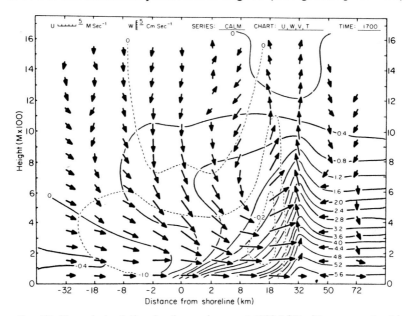

FIG. 68. Numerical solution for the sea breeze at 1700 LST with zero geostrophic wind (vectors give landward and vertical circulation; solid lines are the temperature changes from 0800 LST; dashed lines are the velocity components in meters/sec into the figure) [245].

is quite similar to that of the schematic diagram, Fig. 65. In the case of an off-land geostrophic wind (Fig. 69), a mesoscale cold front develops. Finally, for an off-water geostrophic wind (Fig. 70), the flow is only slightly disturbed in the transitional zone.

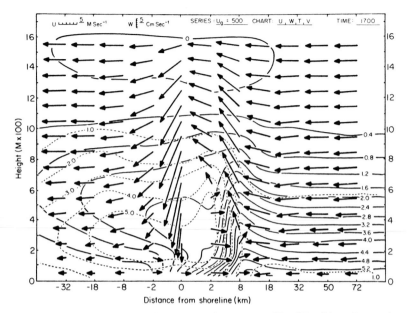

FIG. 69. Numerical solution for the sea breeze at 1700 LST with a 5-meter/sec off-land geostrophic wind [245].

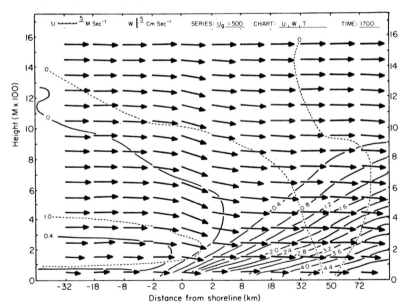

FIG. 70. Numerical solution for the sea breeze at 1700 LST with a 5-meter/sec off-water geostrophic wind [245].

19.5. Humidity Profiles Associated with Sea Breezes

Sharp gradients in water vapor cause anomalous propagation of radio and radar waves. The three-dimensional structure of humidity in the sea-breeze cell is therefore important in radiometeorology, although not well documented.

The classical study is by Hatcher and Sawyer [248] based on aircraft observations in the vicinity of Madras, India. A typical cross section is given in Fig. 71 [249]. The dry "tongue" aloft over the sea is associated with subsiding air.

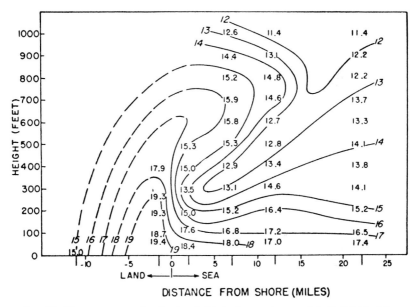

FIG. 71. Vapor pressure distribution (mb) during sea breeze, July 5, 1944 in Massachusetts Bay with an offshore geostrophic wind [249].

19.6. The Micrometeorology of Small Islands and Lakes

A small island is a very local heat source during the daytime. When the geostrophic wind is light, therefore, a convection cell develops with air flowing inward at low levels and rising over the center of the island. A single cumulus cloud will sometimes appear in an otherwise clear sky.

As geostrophic winds increase, the thermal characteristics of the island become less important but the rise of land may introduce aerodynamic effects. Stern and Malkus [250] have obtained numerical solutions predicting the influence of a heated island on the general wind flow.

In the case of a small lake, the circulation cell is in the reverse sense to that of an island. There is air subsidence over the water when convection is present above the surrounding land.

The energy balance of a lake is of considerable hydrological importance. When air blows across a small body of water, two internal boundary layers develop as illustrated schematically in Fig. 72. A few

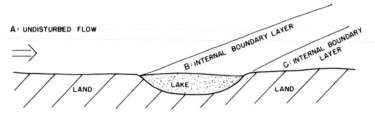

FIG. 72. Internal boundary layers that develop as air moves across a lake.

hundred yards inland from the downwind shore, the two interfaces may lie above each other.

In order to obtain meaningful estimates of evaporation from the lake using vertical profiles or eddy correlations, the sensors must be located in internal boundary layer *B* of Fig. 72. A coarse estimate of the thickness of this layer can be obtained from the theories of Chapter 12.

Water temperatures in small lakes do not show large horizontal variations. The methods of Chapter 18 can therefore be used to obtain estimates of the heat storage within the lake.

20. The Air in Valleys

20.1. The Importance of Valley Meteorology

Most theoretical results in micrometeorology apply to an idealized "infinite plane" or to a simplified transitional zone. However, much of the world is hilly or mountainous, and population is concentrated in valleys rather than over flat terrain. So-called level country is often "gently rolling," permitting valley effects to develop under certain conditions. Many crops are planted on sloping ground.

An infinite plane is very difficult to find in nature. Even in the flattest parts of the world, cities are built along rivers with associated shallow slopes, which may behave as valleys when the geostrophic wind is light. Slope and valley meteorology is therefore of considerable economic importance.

Wind flows in large valleys have been studied in many countries; the scale is such that the Coriolis force (due to the rotation of the earth) is at times in the equations of motion. However, valley flows also develop on the microscale, e.g., in a city street (Chapter 21) or in a forest clearing (Chapter 22). Experimental studies on the microscale are unfortunately not common.

The meteorologically important properties of valleys include:

1. Orientation to the geostrophic wind.
2. *Aspect*, i.e., angle of incidence of the sun's rays. A north-south valley will have quite a different radiation balance from that of one lying in an east-west direction.
3. The geometrical dimensions of the valley, including length, width, depth, slope of the sides, slope of the valley floor, and the presence of bends or constrictions.

20.2. Valley Influences during Strong Geostrophic Winds

When a strong geostrophic wind is blowing in a direction more or less parallel to the valley, there is a funnel effect. Winds are often

189

stronger than over level country, particularly at points where the valley narrows or its sides become steeper.

When the geostrophic wind is blowing in a direction at right angles to the valley, complex flow patterns develop that depend on local geometry. Wind speed is usually less than over level country but there are regions of intense turbulence. Aerodynamic downwash along slopes has been identified as a cause of some aircraft accidents. Figure 73 [251] is a schematic representation of downwash.

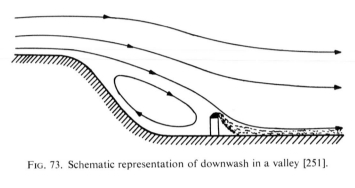

Fig. 73. Schematic representation of downwash in a valley [251].

20.3. The Radiation Balance in Hilly Country

When part of a valley is sunny while the remainder is shaded, two different energy balances must exist. A factor that magnifies the effect is aspect; the energy received on a surface normal to the sun's rays is greater than that received on a horizontal surface. Just after sunrise, for example, a vertical wall facing southeast receives more energy than any other part of the landscape.

Geiger [13] includes an excellent summary of theoretical and experimental studies on the effect of aspect on solar radiation flux. Valleys may be classified according to their orientation and resulting radiation balances. In the case of a north-south valley, the west slope is sunny in the morning, shaded in the afternoon. For an east-west valley, on the other hand, there is no variation near the equator but large differences occur in polar regions with the south slope being almost permanently shaded. Fortunately, the ratio of diffuse to direct solar radiation increases in the Arctic, reducing the magnitude of differences in the energy received by opposite slopes.

There do not appear to be any published investigations of the net radiation balance Q_n of a valley. This is indeed surprising, since net radiation differences constitute a primary "forcing function" for valley flows.

20.4. Local Wind Flows in Valleys during Light Geostrophic Winds

Despite a large number of experimental studies, there is as yet no satisfactory physical model for the wind patterns observed in valleys during light geostrophic winds. It is well known that a wind rose from a valley station shows a preponderance of directions along the length of the valley. That is about the only generality that can be made.

Consider first a slope connecting two large plateaus. At night the slope air cools near the ground and becomes heavier than air at the same level over the lower plateau. Local pressure differences result in a drainage of air down the slope (called a *katabatic* wind) with a weak return circulation aloft.

When the upper plateau is a glacier or when the slope is in the shade, the katabatic wind may occur both day and night. However, when a slope is warmed appreciably by insolation, winds reverse direction (an *anabatic* wind). The anabatic flow is rare in winter when the ground is snow-covered.

In the case of deep, straight canyons or fjords, valley flows are "relatively" simple. At night there is cold air drainage down the valley as well as down the slopes. The slope wind is deflected down the valley to merge with the main flow. If the depth of the valley is D (Fig. 74), wind speed is at a maximum at a height of about $D/2$; the compensating up-valley wind begins at a level slightly below D but is often weak or absent because of interaction with the geostrophic wind.

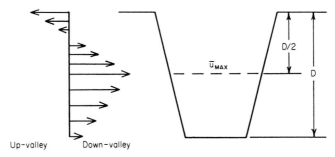

Fig. 74. Schematic representation of valley flows at night during light geostrophic winds.

In the daytime there is an anabatic flow up the valley and up sunny slopes; the pattern shown in Fig. 74 is then reversed.

In valleys of more modest dimensions, particularly when the floor does not slope appreciably, circulations are unfortunately not so clearly defined, although of great practical importance to micrometeorologists. It is unlikely that a "typical" small valley could be found; however, some characteristics that have been observed at particular locations are as follows:

1. *The down-valley direction:* The down-valley direction for cold air drainage is not necessarily the same as for river flow. If the valley narrows in the down-river direction, cold air may flow in the up-river direction toward where the valley broadens. The geostrophic wind also has an effect and cold air drainage may flow in opposite directions on different occasions.

2. *Cross-valley slope winds:* The air over a slope is at a different temperature than the air at the same elevation above the valley floor or (during the day) over the opposite slope. It might be thought that slope winds would be more common during the day than at night but the reverse is actually true. There are few daytime occurrences with the exception of the shaded side of an isolated hill or mountain rising above a flat plain. Daytime superadiabatic lapse rates induce rapid momentum exchanges and tend to obliterate local circulations. At night, on the other hand, inversion conditions diminish the coupling between geostrophic and surface wind flows. Nocturnal slope winds have been observed at a number of locations, e.g., Edmonton, Canada where the valley depth is only 50 meters [252]. Klassen observed the smoke from an ignited oily cloth at Edmonton and found that at the crest of the valley the depth of the slope wind was only 2.5 meters; about 10 meters down into the valley, the depth of the slope wind was even less and some of the smoke was moving parallel to, rather than across, the valley.

3. *Intermittency:* Valley winds, to some extent, and slope winds invariably exhibit intermittency, and it is doubtful whether steady state conditions are maintained for any length of time. Drainage winds occur usually as discrete surges rather than continuously. Buettner and Thyer [253] made balloon observations of wind at night in a deep canyon at 5-min intervals for $2\frac{1}{2}$ hr. Many irregularities were found, the most striking of which was a complete reversal from down-valley to up-valley flow lasting 20–30 min and occurring simultaneously at two stations 2 km apart.

Intermittency is not well understood. One suggested cause is the blocking effect of obstacles. Cold air accumulates behind a line of trees or a building; when the density of the air exceeds some critical value, the air drains suddenly downhill as if a dam had been opened. Other possible causes of intermittency include gravity waves (internal oscillations or seiches), differences in roughness, slope or height of the two sides of the valley with resulting interactions between slope winds, and finally there is the possibility of conflicts between slope and valley winds; for example, in late afternoon a strong up-valley wind may be blowing but there may be cold air drainage down shaded slopes [254].

4. *The effect of minor topographic features:* Koch [255] has emphasized the importance of minor topographic features. An isolated hollow without a drainage exit will become much colder than a valley.

5. *The effect of a river:* A river in the valley floor introduces an additional heat source or sink. For example, Klassen [252] found that slope winds in the valley at Edmonton, Canada drained down to the river edge, then began to rise in a two-cell fashion.

Davidson [256] obtained wind profiles in a number of Vermont valleys during periods of clear skies. A wind speed maximum was found to occur on most nights at a height from 30 to 300 meters above the ground. Davidson defined h as the height at which the valley wind reached its maximum speed u_m and H as the height above h at which the valley flow first became zero or approached a small constant value. The ratio h/H was studied, and on the basis of 214 profiles it had an average value of about 0.5.

The values of H were also correlated with the 1000-meter wind at Albany, New York, about 80 km away. Although there was considerable scatter, it was clear that H decreased with increasing 1000-meter wind. For example, in one valley the height H fell 400 meters as Albany wind increased from 2 meters/sec to 10 meters/sec. The value of u_m, on the other hand, showed no correlation with Albany wind or any other meteorological parameters that were tested. In a later paper [257] it was found that the wind (averaged over the depth of the valley system) increased approximately as the square root of distance from the head of the valley. A further result was that the decay of the Vermont valley wind after sunrise commenced first at ridge level rather than near the floor. Ayer [258] found a similar result in a deep valley in the state of Washington.

20.5. Temperatures and Pollution in Valleys

A valley is a particularly active region, generating its own wind when little is available on the macroscale. Consequently, unless the valley is blocked at both ends, surface minimum temperatures are not as low as in isolated hollows over level terrain. During the day, maximum temperatures are about the same in a valley as over a plain.

Although valleys are favorable locations for inversions, the intensity of the inversion is not great because of the continual overturning of air. The general belief that valleys are poor locations for industrial chimneys is true but not for the reasons usually given, namely, the intensity of inversions and the almost complete disappearance of turbulence. In actual fact, inversions are relatively weak and nighttime turbulence levels are higher than over open country. The real disadvantages of locating a pollution source in a valley are:

1. Downwash with strong geostrophic-level cross winds.
2. The preferred wind direction frequencies for up-valley or down-valley flow. Hence, on an *annual* basis, ground-level concentrations in the valley are higher than if the wind blew equally from all directions.
3. The downslope drainage winds, which may cause fumigations.
4. The sheltering effect of the valley, which increases the number of hours of microscale control of the weather.

21. The Air over a City

21.1. Introduction

City climate is an example of man-made local weather modification. Four physical mechanisms contribute to the effect:

1. The natural radiation balance is disturbed by changes in the properties of the underlying surface. Vegetation is replaced by large areas of concrete and brick. In some of the older residential districts, on the other hand, a canopy of trees has been carefully nurtured, and the climate resembles that of a forest.
2. Built-up areas are obstacles to the wind, changing the natural flow and turbulence of the air.
3. The water vapor balance in a city is upset by the change from moist to dry surfaces.
4. The city emits heat, water vapor, and pollution to the atmosphere. In addition, automobile traffic is a source of local turbulence.

There are a number of important interactions among the four mechanisms. For example, atmospheric pollution affects the radiation balance and the temperature regime.

The climate of cities has been considered at a recent symposium [259] by Geiger [13] and by Kratzer [260]. Since much of the world's population lives in built-up areas, the subject is of great importance. There is growing evidence, for example, that the urban environment contributes substantially to illness and morbidity, particularly in the case of respiratory diseases.

Published climatic summaries may be used in three ways in the search for urban-country weather relations:

1. A rural and a city observing station may be compared.
2. Long-term records (20 to 50 years) may be analyzed for evidence of an effect of increasing industrialization and population growth.

3. Sunday and weekday climatic averages may be compared; commercial and industrial activity are at a minimum on Sunday.

Landsberg [261] emphasizes two difficulties that arise in the identification of city climate by the above methods. In the first place, there are not many comparative records between city and country. " Generally, such parallel series are available by accident rather than by design. This is the case for the usually observed weather elements at a city station and an outlying airport." Second, the urban climate is frequently complicated by special orographic features such as valleys or lakes. It is then not always possible to isolate urban from other contributing factors.

For these reasons there is increasing interest in special micrometeorological studies of a city, although a complete energy balance estimate has yet to be attempted. A concerted *Project Great Plains* type of study [26] is required.

21.2. The Radiation Balance of a City

Sheppard [262] has reviewed the effect of pollution on the radiation balance. There is a reduction in solar energy reaching the ground but the variation with wavelength is not well established either theoretically or experimentally. Some indication of the magnitude of the depletion of shortwave radiation Q_T is given in a study by Hand [263]. In a 4-year comparison of solar radiation measured in downtown Boston and at Blue Hill Observatory 10 miles away, city values averaged 15% lower than those in the suburbs. Mateer [264] found that solar radiation in Toronto, Canada averaged 3% higher on Sundays than on weekdays. During the heating season (October to April inclusive), the increase was 6%, but was only 0.8% for the remainder of the year. Meetham [265] compared ultraviolet radiation $(0.315-0.335 \mu)$ in the center and on the outskirts of Leicester, England over $2\frac{1}{2}$ years. He concluded that in winter the removal of all smoke would have increased ultraviolet radiation by 30% in Central Leicester; in summer the change would have been only 3%. In this connection, although the ultraviolet is not a large fraction of solar radiation, it is of considerable physiological importance.

The summer-winter differences are partly explained by increased smoke emissions in winter. Another factor is the lower angle of the sun (larger optical air mass). In a theoretical analysis of clear sky observations at Kew, England, Blackwell *et al.* [266] calculated the additional absorption and scattering caused by pollution as a percentage of that produced by a pure atmosphere for different solar elevations (Table XXV).

TABLE XXV. Additional clear-sky absorption and scattering caused by the pollution at Kew, England [266].

Solar elevation, deg	54	37	24	12
Additional absorption, %	6	8	11	16
Additional scattering, %	10	10	8	4

The absorption increases while scattering decreases as the solar elevation changes from 54° to 12°. Table XXV refers to a "mean" atmosphere, so that reductions are larger on very smoky days. Roach [267] measured solar radiative flux divergence and albedo from an aircraft over southern England. He concluded that heating rates in excess of 0.5°C/hr may occur within the haze layer above heavily polluted areas due to solar radiative flux divergence; furthermore, the heating cannot be dissipated by additional long-wave cooling.

The albedo of a city is different from that in the country for several reasons:

1. The changed reflectance characteristics of the surface.
2. The presence of vertical surfaces (walls of buildings), which capture more solar energy than a horizontal surface when the angle of the sun is low, and which cause multiple reflections across city streets.
3. The difference in snow-cover between city and country. Snow is quickly removed from streets and soon becomes covered with soot elsewhere in large cities.

There are few comparisons of urban/rural albedo differences. However, some recent aircraft measurements by the University of Wisconsin group are given in Table XXVI [268]. The east and west side measurements were over suburbs that included open areas. A seasonal variation in albedo differences is evident and is dependent upon snow-cover and foliage.

McCormick and Baulch [269] have used *turbidity* as an index of urban particulate pollution. Turbidity is the contribution to the extinction coefficient of equation (2.3) by scattering and absorption of

TABLE XXVI. Albedo measurements over Madison, Wisconsin [268].

	East side	Downtown	West side
Feb. 21, 1963, snow	18	14	42
Mar. 21, 1963	13	15	21
Apr. 11, 1963	15	16	16
Apr. 14, 1963	15	16	16
May 23, 1963	16	23	

solar energy by dusts and aerosols. Vertical profiles were obtained of the ratio I/I_0 at a wavelength of 0.5 μ, using a Voltz photometer mounted on a helicopter. Some typical "polluted air" measurements over Cincinnati, Ohio are given in Fig. 75; concurrent temperature

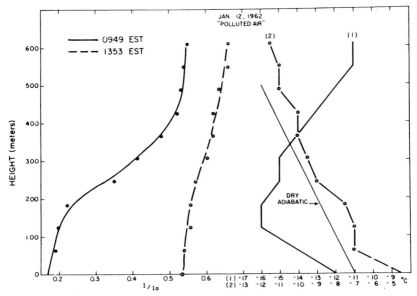

FIG. 75. Variation of transmissivity and temperature with height over Cincinnati, Ohio, Jan. 12, 1962 [269].

profiles are included. The optical transmission improved with the disappearance of the inversion between 0949 and 1353 EST.

The radiation balance of a city has not yet been estimated, although it is of obvious importance. Less short-wave radiation reaches the ground in a city than in the country, but the downward long-wave radiation is greater. In addition, the upward long-wave radiation is less because of many vertical buildings; the walls are exchanging energy with walls across the street as well as with the sky. The relative magnitude of the various components remains to be determined.

21.3. Conductive Heat Flux Q_G

Pavement and concrete buildings have relatively high heat capacities ρc and high conductivities k. The daytime heat storage is therefore greater than in the case of a grass-covered field. A contributing factor is the many glass windows in modern office buildings. Finally, the lack of evaporational cooling increases the energy available for sharing between Q_H and Q_G.

At night, the stored daytime heat is released from buildings and pavements, resulting in higher air temperatures than occur in the country.

Although the behavior of Q_G in a city can be described qualitatively, no quantitative estimates of its magnitude have been suggested.

21.4. The Heat Generated by a City

Kratzer [260] has summarized estimates that have been given for urban energy releases by combustion processes. For the built-up area of Vienna (without parks) the artificial heat supply per annum is one-sixth to one-fourth that provided by direct solar radiation; for Berlin the ratio is one-third. It might be thought, therefore, that the city-country temperature difference would be a maximum in winter when space heating requirements are greatest, particularly if there is snow in the suburbs but not in the center of the city. Such is not the case. It appears that the stronger winter winds increase the advection of heat away from the city. Chandler [270] believes that the heat storage ability of a city is the most important factor, and this is a maximum in summer and early autumn.

The automobile is an important artificial heat source in the downtown "canyons" of large cities. On a calm afternoon in London when traffic was only about one quarter of its maximum density, Gold [271] measured temperatures 2°F higher in Regent Street than in Hanover Square. He estimated that a maximum difference of 8°F was possible.

21.5. City Temperatures

The characteristic warmth of a city is called *the urban heat island.* It is at a maximum at night when skies are clear and winds are light.

Its seasonal maximum is summer–early autumn. A typical case is given in Fig. 76 for London, England [270] based on minimum tem-

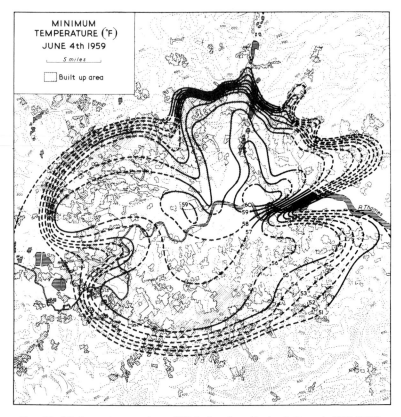

MINIMUM
TEMPERATURE (°F)
JUNE 4th 1959

5 miles

☐ Built up area

FIG. 76. Minimum temperatures (°F) in London, England, June 4, 1959 [270].

peratures for June 4, 1959. Several features are of interest:

1. Winds were light southwesterly, resulting in a slight displacement of the heat island to the northeast.
2. The isotherms were not in the form of concentric smooth circles. Tongues of cold and warm air followed local topographic drainage patterns.
3. The edge of the heat island was very sharp.

Another widely quoted example [272], in which automobile temperature traverses were used in San Francisco, is displayed in Fig. 77. A

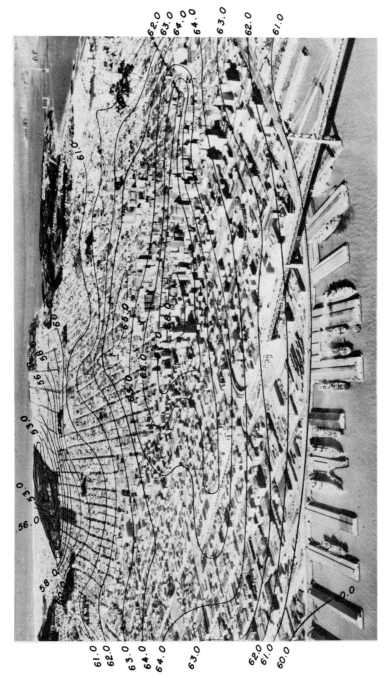

Fig. 77. Isotherm pattern (°F) at 2-meter level in San Francisco, 2320 PST, Apr. 4, 1952 [272].

beneficial effect of the nighttime heat island is to lengthen the frost-free period of the year.

Daytime city temperatures are usually about the same as, or even slightly lower than those in the suburbs [260]; however, Chandler [270] has presented a case in which there was a daytime heat island in London (Fig. 78), although not as strongly developed as nighttime

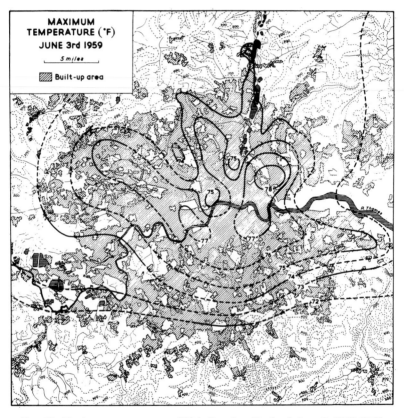

FIG. 78. Maximum temperatures (°F) in London, England, June 3, 1959 [270].

occurrences. Even when no temperature difference exists, the center of a city feels hotter than does the country because of radiant heat transfer from buildings and because of lighter winds.

An example of the vertical temperature structure over a city is given in Fig. 79 [272]. The air over San Francisco was warmer up to a height of 100 ft but cooler aloft than the air over the country.[5] It has also been found that vertical temperature differences from city towers in Louisville, Kentucky [273] and in Montreal, Canada [274] show

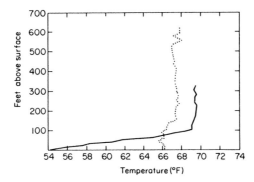

FIG. 79. Vertical profiles in the vicinity of San Francisco at 2210 PST, Mar. 26, 1952; dotted line is for a built-up area; solid line is for an undeveloped area [272].

much lower inversion frequencies than have been reported from rural locations. There are four factors that contribute to the effect:

1. Automobile traffic results in vertical mixing and weakens inversions near the ground.
2. Radiational cooling is intensified at the upper surface of the smoke and carbon dioxide pall over the city.
3. The city heat island induces a convection cell with compensating subsidence in the suburbs.
4. Roofs and treetops may act as an active exchange surface with lapse conditions extending down to street level as in the forest.

Another question of interest is the effect of the size of the city on the intensity of the heat island. It is difficult to find an answer from climatic records because often the so-called rural station used for comparison is drawn into the built-up area over a period of years; as a result, the "apparent" city-country temperature difference decreases. Kratzer [260] concludes that the question cannot be answered but Mitchell [275] finds a significant correlation between population

[5] Summers (First Canadian Conference on Micrometeorology, Toronto, 1965) has suggested that an internal boundary layer develops over a city at times. When stable air is advected over a built-up area, there is a large increase in surface roughness and an input of heat that causes a mixing layer to form and thicken downwind.

growth and urban warming rate in summer but not in winter in a number of United States cities.

Chandler [270] suggests that the magnitude of the heat island is not a linear function of area or population. It appears that a city is a collection of microclimates; the character of the built-up area immediately surrounding the station may be more important than the size or form of the city. The abrupt change in temperature in Fig. 76 at the edge of the heat island supports this view.

21.6. The Humidity in a City

In downtown areas, natural sources of evaporation and transpiration have been removed. The water vapor content of city air is therefore less in summer, as evidenced by lower frequencies of dense fog than in nearby countryside.

In winter, the situation is often reversed because of the emission of water vapor by combustion processes. Robertson [276] has related the occurrence of low temperature fog at Edmonton, Canada to the release of moisture from the combustion of natural gas. Kratzer [260] has compiled many frequency distributions of humidity, fog, cloudiness, and precipitation in cities. Among other things, there is convincing evidence for a weekly cycle that must be related to human activities.

21.7. Winds in a City

During periods of relatively calm weather, the city heat island induces a circulation cell; country air drifts in at low levels toward the center of the city. Buildings often block the advance at street level, although there is a flow above the roof tops. The circulation is therefore not particularly helpful in improving air quality.

Because of local terrain effects, the circulation pattern is never simple. In actual fact, there has been very little experimental evidence until recently that such a circulation did indeed exist, and most of the arguments have been indirect [260]. However, Pooler [277] has found an inflow of air toward the center of Louisville, Kentucky during light geostrophic winds.

Gales are of importance in engineering studies of tall buildings. Davenport [278] believes that city structures have been overdesigned for wind loadings because most climatic data come from exposed airport locations. Another fact to be considered is that dynamic effects are often more important than static pressures; very little information is available concerning gust accelerations or resonance of the wind with the natural period of vibration of the structure (usually less than 1 min). Measurements of strong-wind profiles and associated turbulence spectra in built-up areas are therefore of considerable economic value although not yet available.

21.8. The Effect of Parks and Greenbelts

Frederick [279] has studied the effect of tree leaves in reducing wind speeds, using data from 32 anemometers located at a height of 10 meters in Nashville, Tennessee. Some instruments had relatively open exposures while others were within a few hundred feet of numerous trees. After removing the effect of the generally stronger geostrophic winds that prevail in winter, Frederick found that there was a large seasonal change due to defoliation. For an older residential area containing many well-spaced mature deciduous trees, the increase in wind speed from summer to winter was as much as 40% greater than it was at an exposure nearly devoid of trees.

Trees and hedges have a beneficial effect as filters of smoke and gases. Parks and smokeless zones have the additional advantage of providing space for dilution of pollution from neighboring areas.

Wainwright and Wilson [280] measured concentrations of sulfur dioxide in Hyde Park and Kensington Gardens, London (an area of about 1 square mile). A portable sampler permitted measurements along a line through the middle of the park in the direction of the wind. When wind speeds were less than 3 knots, the SO_2 concentration decreased from the edge but then increased again (Fig. 80a) to about 80% of its initial value at the downwind edge of the park. For stronger winds, the concentration decreased steadily downwind (Fig. 80b) in an exponential fashion:

(21.1) $$R = P_d/P_0 = e^{-Ad}$$

where P_0 is the concentration at upwind edge of park, P_d is the concentration at distance d downwind, and A is an empirical constant. Values of the constant obtained for 18 runs were found to have no correlation with a wind speed measured 2 miles away but were significantly correlated with vertical temperature differences from towers at Crystal Palace (6 miles away) and at Cardington (44 miles away).

The dilution rates are much higher than any reasonable diffusion

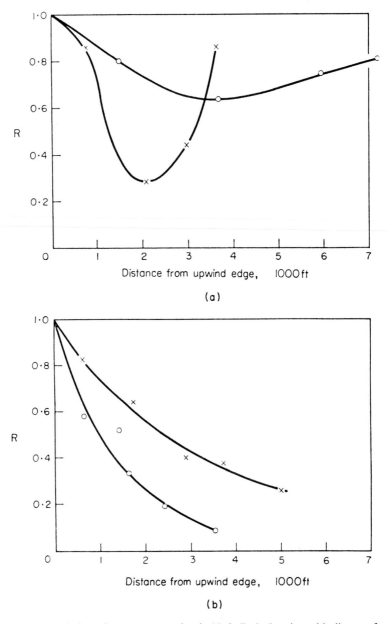

Fig. 80. Variation of SO₂ concentration in Hyde Park, London with distance from upwind edge: (a) winds less than 3 knots, (b) winds greater than 3 knots. The ordinate R is the ratio of concentrations at various downwind distances to that at the upwind edge of the park. Dots and crosses indicate separate trials [280].

model would suggest. Wainwright and Wilson speculate that a current of cleaner air is descending into the park as part of a circulation cell with the surrounding streets and buildings.

A local circulation between parks and built-up areas was also mentioned by Kratzer [260] and by Whiten [281]. Miniature valley circulations may be induced in city streets, for example, when the buildings facing south are in the sun while those facing north are shaded. Chandler [270] therefore is certainly correct when he speaks of a city as a collection of microclimates.

21.9. Conclusion

A knowledge of the climate of cities has many important applications, particularly in bioclimatology, the study of the relations of climate and life. For example, in cities where pollution levels are becoming alarmingly high, it may be necessary to select new areas for industrial zoning on the basis of meteorological factors. In this connection, the *prevailing* wind direction may not be the best criterion; serious smog conditions may be associated with quite a different wind direction.

The prediction of pollution levels from multiple city sources can be made with reasonable accuracy using presently available models [282, 283] in the absence of complicated terrain features. The parameter urgently needed to successfully apply the models is a *source inventory*, an accurate description by city blocks of source strengths and heights, as well as variations by time of day and by season. Studies of this type are developing (e.g., Rosano [284]), but much more needs to be done.

Source inventories are of importance not only in air pollution prediction but also in estimates of the energy balance of a city. There is a very real need for a detailed description of the heat, water vapor, and pollution source strengths and emission heights throughout the urban area. Only then will it be possible to arrive at some reasonable estimates of the energy budget of a city.

Useful advice on the design of chimneys for small urban industrial installations is given in ref. [285].

22. The Modification of Local Weather

22.1. Introduction

Local weather patterns are strongly influenced by the physical properties of the underlying surface. Albedo, the average height of roughness elements, and the availability of water are all-important in determining the energy balance. The ease with which the balance may be disturbed must be evident.

There have been many experimental studies of local weather modification, particularly in connection with frost protection, shelter belts, and forest regeneration. Some useful references include [9, 13, 286, 287]. There are two "self-evident" truths that have been emphasized by many writers:

1. A change in the underlying surface modifies local weather. For example, a climatic record made prior to the construction of an airport is not necessarily indicative of fog and cloudiness frequencies after the runways are built.
2. Conscious weather modifications should not be attempted without a study of the implications of a change. For example, if a shelter belt is grown as protection from strong winds, the incidence of frost may be increased.

22.2. Changes in the Radiation Balance

Major changes in the energy balance can be achieved by modifying the surface albedo. Experimental verification was given as early as 1934 by Ramdas and associates in India. Some of the results are summarized in Table XXVII [288].

Table XXVII displays soil temperatures at Poona, India two weeks after the ground was covered with a thin layer of chalk. The increased albedo of the whitened surface has resulted in a significant lowering of soil temperatures. The mean daily evaporation was 0.56 and 0.29 in. from the control and from the chalk-covered soils, respectively.

TABLE XXVII. The effect of covering bare soil with chalk [288].

	Mean daily maximum temperature, °C	
Depth in soil in cm	Control plot	Plot treated with chalk
0	50.1	31.9
5	31.3	24.1
10	26.1	21.4
15	24.0	20.9
20	23.7	21.4
30	24.7	22.8

Another experiment showing the effect of an albedo change has been reported by Stearns and Lettau [150]. Bushel baskets were placed on Lake Mendota on March 23, 1963. Skies were clear, air temperatures were well above freezing, and an inversion existed over the melting ice surface. The modification consisted of 210 black and 210 white baskets positioned in adjacent areas; each area was 20 meters wide and 21 meters long. Some results are given in Table XXVIII.

TABLE XXVIII. Bushel basket experiment, March 23, 1963; measuring point 18 meters downwind from leading edge [150].

	Black baskets		White baskets	
$z_2 - z_1$, cm	$T_2 - T_1$, °C	Ri	$T_2 - T_1$, °C	Ri
160–80	0.22	0.005	0.68	0.014
80–40	−0.23	−0.005	0.40	0.005
40–20	−0.21	−0.006	0.09	0.004

The values of $(Q_T - Q_R)$ were estimated to be about 0.068 and 0.028 ly/min for the black and white fields, respectively. The effect on vertical temperature differences and on the Richardson number is demonstrated.

A practical application of albedo change is the acceleration of spring breakup of ice by sprinkling a thin layer of dust on the surface.

The principle has been applied for many years in the U.S.S.R. but only recently has there been interest in other countries. Williams and Gold [289] have reviewed the published literature.

Dusting is more effective in Arctic than in Temperate Zones because the melting period is in June-July in polar regions rather than in March-April; a glance at Fig. 2 will show that the daily solar radiation totals are greater in the summer in the Arctic than in the spring in more southerly regions. Williams and Gold estimated the net radiation increase by dusting during the melting period as 90 ly/day in southern Canada and 210 ly/day in the Arctic.

Part of the increase in net radiation is used in raising the surface temperature to the freezing point, thus increasing long-wave radiation $Q_{L\uparrow}$. However, the loss is apparently small. It is also of interest to know the lowest average air temperature for which dusting produces significant melt; a figure of $-10°C$ in the Arctic, with slightly higher values farther south, has been suggested.

The dusting material should have low albedo and high conductivity. If too much is applied, a thin insulating layer may be formed, reducing the heat flux into the ice. Because of its high albedo, salt does not significantly increase the melting rate unless very large quantities are used (750 gm/m² of salt melt less than 1.5 cm of ice at $-4°C$.)

Other ways in which the radiation balance has been modified are through the use of smoke in frost protection and by the shading effect of shelter belts. Although the principle of smoke screens in reducing radiation losses is sound, results have been generally disappointing; in addition, the oily film that may form on fruits and vegetables limits the economic usefulness of the technique.

Shading by a wind-break is only local in character. Although both the long- and short-wave components of radiation are reduced, the effect is only significant for a horizontal distance of from 1 to 2 tree heights [287]; this, of course, may be a very important region in forest regeneration studies.

22.3 Changes in Soil Heat Flux Q_G

It has long been known that a ploughed field is more likely to have frost than compacted soil. The conductivity and resultant heat flow are reduced by the presence of air pockets. For rather similar reasons, a garage roof will often be the only surface in an area displaying frost. Brooks and Schultz [290] emphasize that the frost hazard can be greater in orchards than over open sandy soil in the same environment; the fruit cannot draw a continuing supply of heat from below as in the case of a soil surface. As a final example, covering soil with straw to reduce water losses will increase the diurnal temperature range of the upper surface and the likelihood of frost.

TABLE XXIX. The effect of wetting a soil with the equivalent of one-half inch of rain [288].

	Maximum temperature, °C	
Depth, cm	Control plot	Wetted plot
0	66.4	55.3
0.5	61.0	46.2
2	50.8	40.1
5	41.6	37.2
10	36.9	33.8
15	34.4	32.4
20	33.3	32.9
30	33.5	33.3

Irrigation of a field increases bulk soil conductivity. Table XXIX shows the effect of morning wetting of a soil surface in Poona, India with the equivalent of one-half inch of rain [288]. The maximum afternoon surface temperature was 11°C cooler than in the control plot. Part of the difference is due to the increase in soil conductivity while the remainder is caused by evaporational cooling.

22.4. Changes in Humidity

Airport fog may be cleared by heating the air with lines of oil burners along the runways. Some attempts have also been made to dissipate fog or cloud by releasing Dry Ice (solidified CO_2) or a spray of water from an aircraft. Belyaev and Pavlova [291] were able to change the energy balance in the city of Aktyubinsk in the Soviet Union on the evening of January 11, 1960 by seeding the clouds with Dry Ice. Whereas the surrounding countryside remained cloudy during the night, an area of clear skies, 150 km in diameter, covered the city; at 2300 hr, the temperature was 10°C lower in Aktyubinsk than it was in the rural areas.

Hydrologists are greatly interested in reducing evaporation from open water surfaces in arid zones. Certain long-chain fatty alcohols can be spread on a lake to form monomolecular films at the interface. A decrease in evaporation rates is supposed to result, but there is as yet no convincing evidence to support or to refute the claim. Jarvis [292] has shown that the emissivity is not affected but there may be reductions in the drag coefficient [229].

22.5. Changes in Wind Patterns

Helicopters or wind machines may be used for frost protection. The objective is to promote vertical mixing during inversions, bringing warmer air downward to the plants. The method fails when the frost is

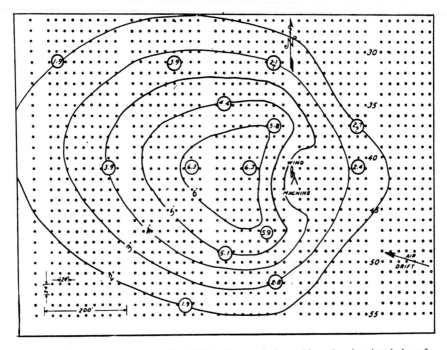

Fig. 81. Orchard response in California to wind machine showing isopleths of wind speed in mph. The little circles are locations of orange trees [293].

caused by a cold wave with positive lapse rates. Studies in California [293] and in Australia [294] have shown that wind machines are operationally useful (see Fig. 81).

The shelter belt is widely used to reduce soil erosion and physical wind damage to crops. The decrease in vertical exchange processes diminishes evapotranspiration and increases agricultural yields. Impressive ecological evidence of the beneficial effects of shelter belts is contained in ref. [287].

The best shelter belt is not too dense but permits some flow of air below the canopy. There are two reasons for this:

1. An impenetrable obstacle promotes downwash during strong winds, increasing vertical fluxes (except in the immediate vicinity of the barrier). When part of the air flow is through the trees, the chances of downwash are reduced. The average wind speeds in the wake of shelter belts have been measured [295]. The 50% reduction isopleth (Fig. 82) is at $7H$, $12H$, and $14H$ for the thick, moderate, and thin belts, respectively, where H is average tree height. The wind velocity reduction is a maximum when the leaf area index is 30% [296].
2. If a field is enclosed by a thick shelter belt, the chances of frost are increased. Some opportunity for cold air drainage is desirable, particularly if advantage can be taken of sloping ground.

A gap in a shelter belt greatly reduces its effectiveness. Wind channeling occurs and the average wind velocity within an otherwise enclosed field is not greatly different from open prairie. When a shelter-belt network is planned, it is therefore important that roads do not form long straight corridors.

Wind speed reduction is greatest when the wind direction is at right angles to the shelter belt. For winds that blow parallel to the belt, the reduction is only about one-quarter as much [295].

Lapse rate has an effect on the efficiency of shelter belts. During superadiabatic conditions, wind recovery in the wake of an obstacle is more rapid because of the stronger vertical flux of momentum. The reverse effect occurs during inversions.

Lapse rate has an effect on wind recovery rate; conversely, a changed wind pattern affects the lapse rate. For example, lower speeds in an enclosed field at night favor inversions. Konstantinov and Vorontsov [296] report that inversions in the lowest 100 meters commenced on the average at 1700 LST in sheltered fields but at 2000 LST over open country. The morning breakup occurred at 0815 and 0700 LST over sheltered and open fields, respectively.

During light winds, local circulations develop between sheltered fields and surrounding trees, an effect that has been demonstrated with neutral-lift balloons. As a result, tree spraying is not successful during the day. The insecticide spreads out over the field. At night, on the other hand, the spray is carried up into the foliage, covering tree stems, branches, and leaves uniformly.

The effect of a series of parallel shelter belts is to reduce the surface wind over the entire area. It therefore requires only a few such belts to change regional values of z_0 and d. A spacing of from 10 to 30

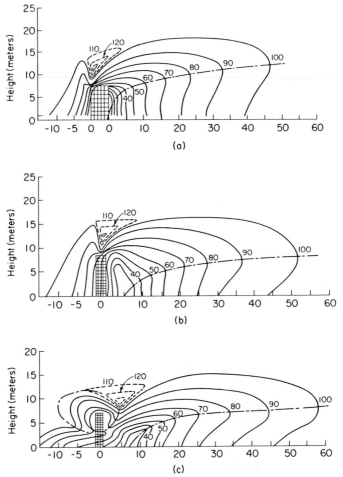

FIG. 82. Percentage wind speed reduction from undisturbed conditions at the same level caused by (*a*) thick, (*b*) moderately wooded, and (*c*) thin forest belt at Kirov, U.S.S.R.; abscissae are multiples of forest belt height [295].

times the tree's height appears desirable for shelter belts, the exact figure depending upon the type of soil [287].

Shelter belts increase soil moisture and cause a greater diurnal amplitude in soil and air temperatures. The aspect of the belts and the slope of the land are important, the resulting patterns of temperature and soil moisture being at times complex.

22.6. Other Examples of Micrometeorological Weather Modification

Almost complete environmental control is achieved within heated and air-conditioned buildings. Another example of weather modification is the design of clothing to provide maximum protection against heat, cold, or strong winds. Micrometeorological knowledge is also required in the planning of storage facilities in warehouses and ships to avoid dampness and excessively high or low temperatures. Although details will not be given here, many of the results of previous sections are relevant. The effects of radiation, conduction, convection, evaporation, and condensation must be considered.

Still another application is in the control of insects, pests, and diseases. Many of these are weather-sensitive, maintaining a life cycle in delicate balance with the environment; for example, ragweed pollen grains are just the right mass and size (20 to 30 μ in diameter) to fall to the ground during light winds but to be carried many kilometers by moderate to strong winds. The way in which each individual species of insect, plant, bacteria, or virus is weather-sensitive may be exploited in micrometeorological control. This, of course, requires the cooperation of several scientific disciplines, a recommendation (or plea) that is the most important underlying philosophy of the author.

An artificial climate of interest to gardeners is that within an unheated greenhouse or cold frame. There is an energy gain relative to the environment because long-wave radiation does not pass readily through glass; however, a most important factor is the protection afforded against turbulent heat losses, which otherwise would quickly dissipate the radiative heat advantage. Jackson [297] has presented data obtained during the winter of 1953–54 at Wye College, England. Cold frames provided some protection against frost at night but the minimum temperature 3 in. above the soil in the frame was never greater than 3°C above the outside minimum temperature at the same level and was often less, particularly on cloudy nights. Maximum temperatures, on the other hand, were greatly elevated on sunny days in the frame, particularly during March (as much as 20°C). There was also a horizontal gradient of soil temperature, the highest values occurring on the south side of the cold frames and away from the edge.

List of Symbols

Radiation

a	Extinction coefficient, equation (2.3)
A	Albedo, equation (2.4)
F	Flux of radiation, equation (2.1)
I_0	Solar constant
I_h	Solar radiation falling on a horizontal surface at the top of the atmosphere
m	Optical air mass, equation (2.3)
Q_{AS}	Rate of short-wave radiation absorption by snow or ice, equation (15.1)
$Q_{L\downarrow}$	Downward long-wave radiation, equation (1.1)
$Q_{L\uparrow}$	Upward long-wave radiation, equation (1.1)
Q_N	Net all-wave radiation, equation (1.1)
Q_R	Reflected short-wave radiation, equation (1.1)
Q_T	Short-wave radiation from sun and sky, equation (1.1)
α	The constant in equation (2.2)
ε	Emissivity
λ_m	Wavelength of maximum energy, equation (2.2)
σ	The constant in equation (2.1)

Below the Interface

c	Specific heat of soil, equation (5.7)
c_d	Specific heat of dry soil, equation (5.7)
c_w	Specific heat of water, equation (5.7)
C	Soil moisture capillary conductivity, equation (5.8)
k	Thermal conductivity, equation (5.1)
K_G	Thermal soil diffusivity, equation (5.6)
K_S	Soil moisture diffusivity, equation (5.8)
M	Integrated horizontal rate of water-mass transport, equation (18.6)
Q_G	Vertical heat flux, equation (1.1)
Q_S	Rate of heat storage, equations (15.1) and (18.5)
Q_{AS}	Rate of short-wave radiation absorption by snow or ice, equation (15.1)
Q_{IW}, Q_{IS}	Heat exchange rates through an ice-water and an ice-snow interface, respectively
t	Time
T	Temperature

List of Symbols

W	Soil moisture, equation (4.1)
W_K	Field capacity soil moisture, equation (4.1)
z	Vertical distance increasing positively downward
ρ	Bulk density of soil

At or near the interface

C_D	Drag coefficient
d	Zero-plane displacement height, equation (7.12)
D	Resistance length, equation (16.10)
E	Rate of evaporation, equation (1.2)
E_p	Rate of potential evaporation
h_1	Equivalent thickness of a layer of living wood equal to the forest biomass, equation (17.1)
H	Ablation rate, equation (15.6)
L	Latent heat of evaporation, equation (1.2)
LAI	Leaf area index, equation (16.1)
Q_E	Latent heat flux, equation (1.1)
Q_H	Sensible heat flux to the air, equation (1.1)
r_S	Resistance, equation (16.8)
R	Bowen's ratio, equation (1.3)
T	Surface temperature, equation (3.1)
T^*	Apparent surface radiative temperature, equation (3.2)
V	Transfer velocity, equation (16.9)
z_0	Roughness length, equation (7.11)
τ_0	Surface shearing stress

Above the Interface

(x,y,z)	Orthogonal coordinate system with x in the direction of the mean wind and z in the vertical
$\bar{u}$	Mean wind speed
$\bar{u} + u', v', w'$	Instantaneous values of the wind in the x,y,z-directions, respectively
A	A measure of the relative efficiency of buoyant and shear forces in producing turbulent energy (see Chapter 9); also, a reference area, equation (13.13); also, an empirical constant in equation (21.3)
B	Swinbank's quasi-constant, equation (18.8)
c_1, c_2	Empirical constant in equation (9.13)
c_p	Specific heat at constant pressure
C	Inertial subrange universal constant, equation (8.6)
C_1	Structure function universal constant, equation (8.9)
$Co(n)$	Cospectrum, equation (8.21)
$CH(n)$	Coherence, equation (8.22)
d	Distance downwind in park, equation (21.1)
D	Depth of valley (Chapter 20)
De	Deacon number of wind profiles, equation (7.14)
DE	Deacon number for temperature profiles, equation (7.15)
e	Water vapor pressure, equation (6.2)
f	Dimensionless frequency (Chapter 8)
$F(n)$	Normalized value of $S(n)$, equation (9.3)

g	Acceleration due to gravity
$G(K, \nu)$	Dimensionless function in the region of local isotropy, equation (8.5)
h	Empirical constant in equation (9.9); also a height in various contexts
H	A height in various contexts
k	von Kármán's constant
K	Wave number, equation (8.2); also, a diffusion parameter in equation (13.13)
K_E	Water vapor diffusivity, equation (10.1)
K_H	Thermal diffusivity, equation (9.4)
K_m	Momentum diffusivity, equation (7.6)
K_x, K_y, K_z	Pollution diffusivities in x,y,z-directions, respectively, equation (13.1)
l_1	Length scale of turbulence equation (8.18)
l_2	Time scale of turbulence, equation (8.19)
$\mathcal{L}$	Monin-Obukhov length, equations (7.18) and (9.1)
n	Frequency, equation (8.1); also, Sutton's n, equation (13.8)
p	Air pressure, equation (6.2); also, power law wind profile exponent, equation (7.21)
P_0, P_d	Concentration of pollution at upwind edge and at a downwind distance d in a park, respectively, equation (21.1)
q	Mean value of specific humidity
$q + q'$	Instantaneous value of specific humidity
q_s	Specific humidity of saturated air
Q	Source strength of pollution, equation (13.4)
Q_H	Sensible heat flux, equation (1.1)
$Q(n)$	Quadrature spectrum, equation (8.22)
Re	Reynolds number, equation (7.1)
Ri	Richardson number, equation (9.5)
Rf	Richardson flux number, equation (9.3)
Rf_{crit}	Critical Richardson flux number (Section 9)
$R(x)$	Autocorrelation function, equation (8.13)
s	Smoothing time, equation (8.10)
s_y^2, s_z^2	Variances of plume dispersion in the y,z-directions, respectively, equation (13.4)
S	Structure function, equation (8.8)
$S(n)$	One-dimensional component of spectral density, equation (8.3)
SR	Stability ratio, equation (9.7)
St	Strouhal number (Chapter 11)
T	Mean temperature; also, length of record in equation (8.12)
$T + T'$	Instantaneous value of the temperature
u_*	Friction velocity, equation (7.5)
X	Mixing ratio, equation (6.2)
L, E	(Subscripts) Lagrangian and Eulerian reference frames, equation (8.25)
α	Monin-Obukhov constant, equation (7.19); also, the constant in equation (13.12)
β	Deacon's exponent, equation (7.13); also, the ratio of Lagrangian to Eulerian time scales, equation (8.25)
Γ	Dry adiabatic lapse rate, equation (6.1)
ε	Rate of viscous dissipation, equation (7.7)

ε_T	Rate of viscous dissipation of turbulent temperature fluctuations, equation (9.18)
θ	Potential temperature, equation (6.1)
λ	Wavelength, equation (8.1)
μ	Dynamic viscosity, equation (7.3)
ν	Kinematic viscosity, equations (7.1) and (7.4)
ξ	Lagrangian time, equation (8.25)
ρ	Air density
σ_A, σ_E	Standard deviations of azimuth and elevation angles, respectively, equations (13.10) and (13.11)
$\sigma_y{}^2, \sigma_z{}^2$	Variances of Lagrangian particle dispersion in the y,z-directions, respectively, equation (13.5)
τ	Shearing stress, equation (7.3)
ϕ	Monin-Obukhov function, equation (7.16)
χ	Ground-level concentration of pollution, equation (13.4)

References

1. Bowen, I. S. (1926). The ratio of heat losses by conduction and by evaporation from any water surface. *Phys. Rev.* **27**, 779–787.
2. Allen, C. W. (1958). Solar radiation. *Quart. J. Roy. Meteorol. Soc.*, **84**, 307–318.
3. Mateer, C. L., and Godson, W. L. (1959). A nomogram for the determination of solar altitude and azimuth. *Monthly Weather Rev.* **87**, 15–18.
4. List, R. J., ed. (1958). "Smithsonian Meteorological Tables," 527pp. Smithsonian Inst., Washington, D.C.
5. Moon, P. (1940). Proposed standard solar-radiation curves for engineering use. *J. Franklin Inst.* **230**, 583–617.
6. Munn, R. E., and Truhlar, E. J. (1963). The energy budget approach to heat transfer at the surface of the earth. *Transact. Eng. Inst. Canada* **6**, B–7, 1–20.
7. Fritz, S. (1949). Solar radiation on cloudless days. *Heat. Vent.* **46**, 69–74.
8. Jen-Hu-Chang (1958). "Report of Ground Temperature," Vol. 1, 300pp. Blue Hill Meteorol. Obs. (Harvard), Milton, Massachusetts.
9. Budyko, M. I. (1956). "The Heat Balance of the Earth's Surface," 259pp. (English transl. from the Russian by U.S. Dept. of Commerce, Washington, D.C., 1958.)
10. Nicolet, M. (1948). La mésure du rayonnement solaire. *Inst. Roy. Meteorol. Belg.* **21**, 3–37.
11. Latimer, J. R. (1962). Laboratory and field studies of the properties of radiation instruments. Rep. No. Tec-414, 13pp. Meteorol. Branch, Toronto.
12. Bauer, K. G., and Dutton, J. A. (1960). Flight investigation of surface albedo. Tech. Rep. No. 2, 69pp. Dept. Meteorol., Univ. of Wisconsin, Madison, Wisconsin.
13. Geiger, R. (1959). "The Climate Near the Ground," 494pp. (English transl. from the German by Harvard Univ. Press, Cambridge, Massachusetts.)
14. Shaw, J. H. (1954). Infrared studies of the atmosphere. Final Rep., 71pp. Ohio State Univ. Res. Found. Proj. RF-381, Rep. No. 23, AF 19(122)65.
15. Yamamoto, G. (1952). On a radiation chart. *Sci. Rep. Tohoku Univ., Geophys.* **4**, 9–23.
16. Kraus, H. (1963). Der Tagesgang des Energiehaushaltes der bodennahen Luftschicht. *Arch. Meteorol. Geophys. Biokl., B* **12**, 491–515.
17. Robinson, G. D. (1950). Notes on the measurement and estimation of atmospheric radiation. *Quart. J. Roy. Meteorol. Soc.* **76**, 37–51.
18. Funk, J. P. (1960). Measured radiative flux divergence near the ground at night. *Quart. J. Roy. Meteorol. Soc.* **86**, 382–389.
19. Funk, J. P. (1961). A numerical method for the computation of the radiative flux divergence near the ground. *J. Meteorol.* **18**, 388–392.

221

20. Funk J. P. (1964). Comparison of measured and computed radiative flux divergence profiles near the ground. Abst. for presentation at I.A.M.A.P. *Radiation Symp., Leningrad.*, 1964.

21. Hamilton, H. L. Jr. (1964). The use of a sweeping boom mechanism in a study of low-level radiation flux divergence. *Am. Meteorol. Soc. Natl. Conf. Micrometeorol., Salt Lake City*, 1964.

22. Suomi, V. E., Franssila, M., and Islitzer N. F. (1954). An improved net-radiation instrument. *J. Meteorol.* **11**, 276–282.

23. Latimer, J. R. (1963). The accuracy of net radiometers. *In* Symposium on the heat exchange at snow and ice surfaces. Tech. Mem. No. 78, pp. 31–55. Natl. Res. Council, Ottawa.

24. Harrell, W., and Richardson, E. A. (1960). Measurement of sensible air-ground interface temperatures. *Monthly Weather Rev.* **88**, 269–273.

25. Monteith, J. L. and Sceicz, G. (1962). Radiative temperature in the heat balance of natural surfaces. *Quart. J. Roy. Meteorol. Soc.* **88**, 496–507.

26. Lettau, H. H., and Davidson, B., eds. (1957). "Exploring the Atmosphere's First Mile," 2 vol. Pergamon Press, Oxford.

27. Slatyer, R. O., and McIlroy, I. C. (1961). "Practical Microclimatology," 250pp. C.S.I.R.O., Australia and UNESCO.

28. John, P. T. (1961). Evaporation from the top layers of soil. *Indian J. Meteorol. Geophys.* **12**, 590–597.

29. Buettner, K. J. K. (1958). Sorption by the earth surface and a new classification of kata-hydrometeoric processes. *J. Meteorol.* **15**, 155–163.

30. Philip, J. R. (1957). Evaporation, and moisture and heat fields in the soil. *J. Meteorol.* **14**, 354–366.

31. Sarson, P. B. (1960). Exceptional sudden changes of earth temperature. *Meteorol. Mag.* **89**, 201–208.

32. Maruyama, E. (192). Studies on the evaporation from the soil surface. *J. Meteorol. Res. (Japan)* **14**, 1–32 (in Japanese).

33. Molga, M. (1962). "Agricultural Meteorology," 350pp. (English transl. from the Polish by U.S. Dept. of Commerce, Washington D.C., OTS-60-21419.)

34. Penman, H. L. (1962). Weather and crops. *Quart. J. Roy. Meteorol. Soc.* **88**, 209–219.

35. Bonner, J. (1962). The upper limit of crop yield. *Science* **137**, 11–15.

36. Penman, H. L. (1951). The role of vegetation in meteorology, soil mechanics and hydrology. *Brit. J. Appl. Phys.* **2**, 145–151.

37. Gardner, W. R., and Ehlig, C. F. (1963). The influence of soil water on transpiration by plants. *J. Geophys. Res.* **68**, 5719–5724.

38. King, K. M., Tanner, C. B., and Suomi, V. E. (1956). A floating lysimeter and its evaporation recorder. *Transact. Am. Geophys. Un.* **37**, 738–742.

39. Pruitt, W. O., and Angus, D. E. (1960). Large weighing lysimeter for measuring evapotranspiration. *Transact. Am. Soc. Agr. Engrs.* **3**, 13–18.

40. Pelton, W. L. (1961). The use of lysimetric methods to measure evapotranspiration. *In* "Proceedings of Hydrology Symposium No. 2," pp. 106–122. Cat. R32–361/2, Queen's Printer, Ottawa.

41. McIlroy, I. C., and Angus, D. E. (1963). "The Aspendale multiple weighed lysimeter installation." Tech. Paper No. 14, 27pp. C.S.I.R.O. Meteorol. Phys. Australia.

42. Blanc, M. L. (1958). The climatological investigation of soil temperature. Tech. Note No. 20, 18pp. World Meteorol. Org., Geneva.

43. Rider, N. E. (1958). Measurement of evaporation, humidity in the biosphere and soil moisture. Tech. Note No. 21, 49pp. World Meteorol. Org., Geneva.

44. Lettau, H. H. (1954). Improved models of thermal diffusion in the soil. *Transact. Am. Geophys. Un.* **35**, 121–132.

45. Portman, D. J. (1958). Conductivity and length relationships in heat-flow transducer performance. *Transact. Am. Geophys. Un.* **39**, 1089–1094.

46. Philip, J. R. (1961). The theory of heat flux meters. *J. Geophys. Res.* **66**, 571–579.

47. Kaganov, M. A., and Rosenstock, Yu. L. (1961). The measurement of heat flow with a heat flow meter. *Bull. Acad. Sci. U. S. S. R., Geophys. Ser.* **8**, pp. 1174–1178. (English transl. from the Russian by Am. Geophys. Un.)

48. Padmanabhamurty, B., and Subrahmanyam, V. P. (1961). Diurnal and seasonal variations of heat-flow into soil at Waltair. *Indian J. Meteorol. Geophys.* **12**, 261–266.

49. Carson, J. E., and Moses, H. (1963). The annual and diurnal heat-exchange cycles in upper layers of soil. *J. Appl. Meteorol.* **2**, 397–406.

50. Romanova, E. N., and Kaulin, N. Ya. (1960). The methodological problem of measuring the minimum temperature at the surface of the soil. *Proc. Main Geophys. Obs.* **91**, 62–70. (English transl. from the Russian by Am. Meteorol. Soc. T–R–385, AF 19(604)–6113.)

51. Hewson, E. W. (1945). The meteorological control of atmospheric pollution by heavy industry. *Quart. J. Roy. Meteorol. Soc.* **71**, 266–282.

52. Funk, J. P. (1962). Radiative flux divergence on radiation fog. *Quart. J. Roy. Meteorol. Soc.* **88**, 233–249.

53. Rodhe, B. (1962). The effect of turbulence on fog formation. *Tellus* **14**, 49–86.

54. Lettau, H. H. (1961). A generalized mathematical model of the mean-velocity distribution in fully turbulent duct flow. Ann. Rep. No. DA-36–039-SC-80282, pp. 115–142. Dept. Meteorol., Univ. of Wisconsin, Madison, Wisconsin.

55. Langhaar, H. L. (1951). " Dimensional Analysis and Theory of Models," 166pp. Wiley, New York.

56. Sutton, O. G. (1953). "Micrometeorology," 333pp. McGraw-Hill, New York.

57. Lettau, H. H. (1964). A new vorticity-transfer hypothesis of turbulence theory. *J. At. Sc.* **21**, 453–456.

58. Calder, K. L. (1939). A note on the constancy of horizontal turbulent shearing stress in the lower layers of the atmosphere. *Quart. J. Roy. Meteorol. Soc.* **65**, 537–541.

59. Monin, A. S., and Obukhov, A. M. (1954). Basic laws of turbulent mixing in the ground layer of the atmosphere. *Geophys. Inst., Acad. Sci. U. S. S. R.* **24**, No. 151, 163–187. (English transl. from the Russian by Am. Meteorol. Soc. T–R–174, AF 19(604)–1936.)

60. Dalrymple, P. C., Lettau, H. H., and Wollaston, S. H. (1963). South pole micrometeorology program. Rep. No. 20, 94pp. Inst. Polar Studies, Ohio State Univ., Columbus, Ohio (ES-7).

61. Brunt, D. (1941). "Physical and Dynamical Meteorology," 428pp. Cambridge Univ. Press, London and New York.

62. Taylor, R. J. (1952). The dissipation of kinetic energy in the lowest layers of the atmosphere. *Quart. J. Roy. Meteorol. Soc.* **78**, 179–185.

63. Stewart, R. W. (1962). Unpublished comments. *Third Natl. Cong. Can. Branch Roy. Meteorol. Soc., Hamilton, Canada,* 1962.
64. Brooks, F. A. (1963). Physical interpretations of diurnal variations of eddy transfers near the ground. Final Rep. No. DA-36–039-SC-80334, pp. 27–65. Dept. Agr. Eng. and Dept. Irr., Univ. of California, Davis, California.
65. Deacon, E. L. (1949). Vertical diffusion in the lowest layers of the atmosphere. *Quart. J. Roy. Meteorol. Soc.* **75**, 89–103.
66. Lettau, H. H. (1962). Notes on theoretical models of profile structure in the diabatic surface layer. Final Rep. No. DA-36–039-SC-80282, pp. 195–226. Dept. Meteorol., Univ. Wisconsin, Madison, Wisconsin.
67. Rider, N. E. (1954). Eddy diffusion of momentum, water vapour, and heat near the ground. *Phil. Transact. Roy. Soc., A* **246**, 481–501.
68. Lettau, H. H. (1949). Isotropic and non-isotropic turbulence in the atmospheric surface layer. *Geophys. Res. Pap. U.S. AFCRC* No. 1, 86pp.
69. Priestley, C. H. B. (1963). Recent flux and profile measurements in Australia. Abst. in *Proc. 13th Assembly IUGG-IAMAP Berkeley, Calif.,* 1963, p. 178.
70. Taylor, R. J. (1960). Similarity theory in the relation between fluxes and gradients in the lower atmosphere. *Quart. J. Roy. Meteorol. Soc.* **86**, 67–78.
71. Munn, R. E., and Richards, T. L. (1963). The micrometeorology of Douglas Point, Ont. Rep. No. Tec-455, 13pp. Meteorol. Branch, Toronto.
72. DeMarrais, G. A. (1959). Wind-speed profiles at Brookhaven National Laboratory. *J. Meteorol.* **16**, 181–190.
73. Panofsky, H. A., Blackadar, A. K., and McVehil, G. E. (1960). The diabatic wind profile. *Quart. J. Roy. Meteorol. Soc.* **86**, 390–398.
74. Sheppard, P. A. (1963). Momentum and other exchange above a water surface. Abst. in *Proc. 13th Assembly, IUGG-IAMAP Berkeley, Calif.,* 1963, p. 117.
75. Goddard, W. B. (1963). Introductory measurements of shear-stress across rye grass sod. Final Rep. No. DA-36–039-SC-80334, pp. 149–157. Dept. Agr. Eng. and Dept. Irr., Univ. of California, Davis, California.
76. Gurvich, A. S. (1961). Measurement of frictional stresses in the near-ground layer of the atmosphere. *Bull. Acad. Sci. U. S. S. R., Geophys. Ser.* **3** pp. 458–466. (English transl. from the Russian by Am. Geophys. Un.)
77. Stewart, R. W. (1956). A new look at the Reynold's stresses. *Can. J. Phys.* **34**, 722–725.
78. Panofsky, H. A., and Pasquill, F. (1963). The constant of the Kolmogorov law. *Quart. J. Roy. Meteorol. Soc.* **89**, 550–551.
79. Cramer, H. E. (1952). Preliminary results of a program for measuring the structure of turbulent flow near the ground. *Geophys. Res. Pap. U.S. AFCRC* No. 19, 187–205.
80. Frenkiel, F. N. (1964). Experimental study of the microstructure of turbulence by high-speed computing technique. *Am. Meteorol. Soc. Natl. Conf. on Micrometeorol., Salt Lake City,* 1964.
81. Kolmogorov, A. N. (1962). A refinement of previous hypotheses concerning the local structure of turbulence in a viscous incompressible fluid at high Reynolds number. *J. Fl. Mech.* **13**, 82–85.
82. Obukhov, A. M. (1962). Some specific features of atmospheric turbulence. *J. Geophys. Res.* **67**, 3011–3014.

83. Novikov, E. A., and Stewart, R. W. (1964). The intermittency of turbulence and the spectrum of energy dissipation fluctuations. *Bull. Acad. Sci. U. S. S. R., Geophys. Ser.* 3, 408–413. (English transl. from the Russian by Am. Geophys. Un.)

84. Hinze, J. O. (1959). "Turbulence," 586pp. McGraw-Hill, New York.

85. Davenport, A. G. (1961). The spectrum of horizontal gustiness near the ground in high winds. *Quart. J. Roy. Meteorol. Soc.* 87, 194–211.

86. Pasquill, F. (1962). "Atmospheric Diffusion," 297pp. Van Nostrand, Princeton, New Jersey.

87. Kolmogorov, A. N. (1941). The local structure of turbulence. *Compt. rend. Acad. Sci. URSS* 30, 301–305 (in Russian).

88. Grant, H. L., Stewart, R. W., and Moilliet, A. (1962). Turbulence spectra from a tidal channel. *J. Fl. Mech.* 12, 241–268.

89. Zubkovskii, S. L. (1962). Frequency spectra pulsations of the horizontal component of wind velocity in the surface air layer. *Bull. Acad. Sci. U. S. S. R., Geophys. Ser.* 10, pp. 1425–1433. (English transl. from the Russian by Am. Geophys. Un.)

90. Gifford, F. (1959). The interpretation of meteorological spectra and correlations. *J. Meteorol.* 16, 344–346.

91. Obukhov, A. M., and Yaglom, A. M. (1951). The microstructure of turbulent flow. *Prikl. Matem. i Mekh.* 15, 3–26 (in Russian).

92. Takeuchi, K. (1962). On the nondimensional rate of dissipation of turbulent energy in the surface boundary layer. *J. Meteorol. Soc. Japan, Ser.* 2 40, 127–135.

93. Priestley, C. H. B. (1959). " Turbulent Transfer in the Lower Atmosphere," 130pp. Univ. of Chicago Press, Chicago, Illinois.

94. Ball, F. K. (1961). Viscous dissipation in the atmosphere. *J. Meteorol.* 18, 553–555.

95. Taylor, G. I. (1938). The spectrum of turbulence. *Proc. Roy. Soc. A* 164, 476–490.

96. Panofsky, H. A., Cramer, H. E., and Rao, V. R. K. (1958). The relation between Eulerian time and space spectra. *Quart. J. Roy. Meteorol. Soc.* 84, 270–273.

97. Panofsky, H. A., and Brier, G. W. (1958). "Some Applications of Statistics to Meteorology," 224pp. College of Mineral Industries, Penn. State Univ., University Park, Pennsylvania.

98. Swinbank, W. C. (1955). An experimental study of eddy transports in the lower atmosphere. Tech. Paper No. 2, 30pp. C.S.I.R.O. Meteorol. Phys. Australia.

99. Panofsky, H. A. (1963). Determination of stress from wind and temperature measurements. *Quart. J. Roy. Meteorol. Soc.* 89, 85–94.

100. Pasquill, F. (1963). The determination of eddy diffusivity from measurements of turbulent energy. *Quart. J. Roy. Meteorol. Soc.* 89, 95–106.

101. MacCready, P. B., and Jex, H. R. (1964). Turbulent energy measurements by vanes. *Quart. J. Roy. Meteorol. Soc.* 90, 198–203.

102. Hay, J. S., and Pasquill, F. (1959). Diffusion from a continuous source in relation to the spectrum and scale of turbulence. *Adv. Geophys.* 6, 345–365.

103. Angell, J. K. (1963). Measurements of Lagrangian and Eulerian properties of turbulence at a height of 2300 ft. *Quart. J. Roy. Meteorol. Soc.* 90, 57–71.

104. Wandel, C. F., and Kofoed-Hansen, O. (1962). On the Eulerian-Lagrangian transform in the statistical theory of turbulence. *J. Geophys. Res.* 67, 3089–3093.

105. Lumley, J. L., and Panofsky, H. A. (1964). "The Structure of Atmospheric Turbulence," 239pp. Wiley, New York.

106. Gill, G. C. (1963). Data validation. Publ. No 79, 23pp. Dept. Meteorol. and Oceanog., Univ. of Michigan, Ann Arbor, Michigan.

107. MacCready, P. B., and Jex, H. R. (1964). Response characteristics and meteorological utilization of propeller and vane wind sensors. *J. Appl. Meteorol.* **3**, 182–193.

108. Richardson, L. F. (1920). The supply of energy from and to atmospheric eddies. *Proc. Roy. Soc.*, *A* **97**, 354–373.

109. Ellison, T. H. (1957). Turbulent transport of heat and momentum from an infinite rough plane. *J. Fl. Mech.* **2**, 463–466.

110. Robinson, G. D. (1959). Vertical motion and the transfer of heat and momentum near the ground. *Adv. Geophys.* **6**, 259–268.

111. Senderikhina, I. L. (1961). On the relationships among the coefficients of turbulent momentum, heat and matter in the surface layer of the atmosphere. *Proc. Main Geophys. Obs.* **121**, 1–23. (English transl. from the Russian by U.S. Dept. of Commerce, 1963, JPRS: 17, 299.)

112. Swinbank, W. C. (1964). The exponential wind profile. *Quart. J. Roy. Meteorol. Soc.* **90**, 119–135.

113. Townsend, A. A. (1962). Natural convection in the earth's boundary layer. *Quart. J. Roy. Meteorol. Soc.* **88**, 51–56.

114. Priestley, C. H. B. (1954). Convection from a large horizontal surface. *Australian J. Phys.* **7**, 176–201.

115. Stewart, R. W. (1959). The problem of diffusion in a stratified fluid. *Adv. Geophys.* **6**, 303–311.

116. Monin, A. S. (1962). Empirical data on turbulence in the surface layer of the atmosphere. *J. Geophys. Res.* **67**, 303–311.

117. Kazansky, A. B., and Monin, A. S. (1956). Turbulence in the inversion layer near the surface. *Izv. Akad. Nauk. SSSR, Ser. Geofiz.* **1**, 79–86 (in Russian).

118. Yamamoto, G. (1959). Theory of turbulent transfer in non-neutral conditions. *J. Meteorol. Soc. Japan* **37**, 60–69.

119. Panofsky, H. A. (1961). An alternative derivation of the diabatic wind profile. *Quart. J. Roy. Meteorol. Soc.* **87**, 109–110.

120. Sellers, W. D. (1962). A simplified derivation of the diabatic wind profile. *J. Atmos. Sci.* **19**, 180–181.

121. Taylor, R. J., and Dyer, A. J. (1958). An instrument for measuring evaporation from natural surfaces. *Nature* **181**, 408–409.

122. Perepelkina, A. V. (1959). On the determination of turbulent flow of heat. *Bull. Acad. Sci. U. S. S. R., Geophys. Ser.* **7**, pp. 1026–1035. (English transl. from the Russian by Am. Geophys. Un.)

123. Gurvich, A. S., and Zwang, L. R. (1960). The spectral composition of turbulent heat flow. *Bull. Acad. Sci. U. S. S. R., Geophys. Ser.* **10**, pp. 1547–1548. (English transl. from the Russian by Am. Geophys. Un.)

124. Dyer, A. J. (1961). Measurements of evaporation and heat transfer in the lower atmosphere by an automatic eddy-correlation technique. *Quart. J. Roy. Meteorol. Soc.* **87**, 401–412.

125. Sternberg, J. (1962). A theory for the viscous sublayer of a turbulent flow. *J. Fl. Mech.* **13**, 241–271.

126. Zwang, L. R. (1960). Measurements of temperature pulse frequency spectra in the surface layer of the atmosphere. *Bull. Acad. Sci. U. S. S. R., Geophys. Ser.* **8**, pp. 1252–1262. (English transl. from the Russian by Am. Geophys. Un.)

127. Taylor, R. J. (1960). A new approach to the measurement of turbulent fluxes in the lower atmosphere. *J. Fl. Mech.* **10**, pp. 449–458.

128. Gaevskaya, G. N., Kondrati'ev, K. Y., and Yakushevskaya, K. E. (1963). Radiative heat flux divergence and heat regime in the lowest layer of the atmosphere. *Arch. Meteorol., Geophys. Biokl., B* **12**, 95–107.

129. Elliott, W. P. (1964). The height variation of vertical heat flux near the ground. *Quart. J. Roy. Meteorol. Soc.* **90**, 260–265.

130. Hickman, K. C. (1963). From drops to boules: the maximum evaporation of water. Abst. in *Proc. 13th General Assembly IUGG-IAMAP Berkeley, Calif.*, 1963, p. 124.

131. Sechrist, F. (1963). Effect of carbon dioxide on evaporation of water. *Nature* **199**, 899–900.

132. Thornthwaite, C. W., and Holzman, B. (1939). The determination of evaporation from land and water surfaces. *Monthly Weather Rev.* **67**, 4–11.

133. Clayton, W. H., and Mistry, P. D. (1963). A study of spatial variations in micrometeorological parameters. Final Rep. No. 214, 63–15F, AF 19(604)–5527, 56pp. Dept. Oceanog. and Meteorol., Texas A. & M., College Station, Texas.

134. Crawford, T. V. (1965). Moisture transfer in free and forced convection. *Quart. J. Roy. Meteorol. Soc.* **91**, 18–27.

135. Taylor, R. J. (1963). The effect of instrumental inertia on measurement of the turbulent flux of water vapour. *Australian J. Appl. Sci.* **14**, 27–37.

136. Pruitt, W. O. (1963). Application of several energy balance and aerodynamic evaporation equations under a wide range of stability. Final Rep. No. DA-36–039-SC-80334, pp. 107–123. Dept. Agr. Eng. and Dept. Irr., Univ. of California, Davis, California.

137. Kaimal, J. C., and Businger, J. A. (1963). A continuous wave sonic anemometer-thermometer. *J. Appl. Meteorol.* **2**, 156–164.

138. Elagina, L. G. (1962). Optical device for measuring the turbulent pulsations of humidity. *Bull. Acad. Sci. U. S. S. R., Geophys. Ser.* **8**, pp. 1100–1107. (English transl. from the Russian by Am. Geophys. Un.)

139. Woodruff, N. P., and Zingg, A. W. (1955). A comparative analysis of wind-tunnel and atmospheric air-flow patterns about single and successive barriers. *Transact. Am. Geophys. Un.* **36**, 203–208.

140. Halitsky, J. (1962). Diffusion of vented gas around buildings. *J. Air Poll. Control Assoc.* **12**, 74–80.

141. Moses, H., and Daubek, H. G. (1961). Errors in wind measurements associated with tower-mounted anemometers. *Bull. Am. Meteorol. Soc.* **42**, 190–194.

142. Rider, N. E. (1960). On the performance of sensitive cup anemometers. *Meteorol. Mag.* **89**, 209–215.

143. Gill, G. C. (1964). Exposure of instruments. *In* "Meteorological Aspects of Air Pollution: Training Course Manual," pp. 143–147. U.S. Public Health Service, Cincinnati, Ohio.

144. Brooks, F. A. (1961). Need for measuring horizontal gradients in determining vertical eddy transfers of heat and moisture. *J. Meteorol.* **18**, 589–596.

145. Inoue, S., Tani, N., Imai, K., and Isobe, S. (1958). The aerodynamic measurement of photosynthesis over a nursery of rice plants. *J. Agr. Meteorol. Japan* **14**, 45–53. (English transl. from the Japanese by Am. Meteorol. Soc. T–J–14, AF 19(604)–6113.)

146. Elliott, W. P. (1958). The growth of the atmospheric internal boundary layer. *Transact. Am. Geophys. Un.* **39**, 1048–1054.
147. Taylor, R. J. (1962). Small-scale advection and the neutral wind profile. *J. Fl. Mech.* **13**, 529–539.
148. Panofsky, H. A., and Townsend, A. A. (1964). Change of terrain roughness and the wind profile. *Quart. J. Roy. Meteorol. Soc.* **90**, 147–155.
149. Kutzbach, J. E. (1961). Investigations of the modification of wind profiles by artificially controlled surface roughness. Ann. Rep. No. DA-36-039-SC-80282, pp. 71–113. Dept. Meteorol., Univ. of Wisconsin, Madison, Wisconsin.
150. Stearns, C. R. and Lettau, H. H. (1963). Report on two wind-profile modification experiments in airflow over the ice of Lake Mendota. Ann. Rep. No. DA-36-039-AMC-00878, pp. 115–138. Dept. Meteorol., Univ. of Wisconsin, Madison, Wisconsin.
151. Portman, D. J. and Brock, F. V. (1962). Analog computer simulation for the advection equation. Second Ann. Rep. No. DA-36-039-SC-80334, pp. 115–127. Dept. Agr. Eng. and Dept Irr., Univ. of California, Davis, California.
152. Philip, J. R. (1959). The theory of local advection. *J. Meteorol.* **16**, 535–547.
153. Rider, N. E. and Philip, J. R. (1960). Advection and evaporation. *Assoc. Int. Hydrol. Sci.* **53**, 421–427.
154. Dyer, A. J., and Pruitt, W. O. (1962). Eddy-flux measurements over a small irrigated area. Second Ann. Rep. No. DA-36-039-SC-80334, pp. 47–52. Dept. Agr. Eng. and Dept. Irr., Univ. of California, Davis, California.
155. Sheppard, P. A. (1962). Properties and processes at the earth's surface in relation to the general circulation of the atmosphere. *Adv. Geophys.* **9**, 77–96.
156. Rider, N. E., Philip, J. R., and Bradley, E. F. (1963). Horizontal transport of heat and moisture—a micrometeorological study. *Quart. J. Roy. Meteorol. Soc.* **89**, 507–531.
157. Dyer, A. J. (1963). The adjustment of profiles and eddy fluxes. *Quart. J. Roy. Meteorol. Soc.* **89**, 276–280.
158. Miyake, M. (1961). Transformation of atmospheric boundary layer. M.A. Thesis 65pp. Dept. Meteorol., Univ. of Washington, Seattle, Washington.
159. Estoque, M. A. (1963). A numerical model of the atmospheric boundary layer. *J. Geophys. Res.* **68**, 1103–1113.
160. Monin, A. S. (1959). General survey of atmospheric diffusion. *Adv. Geophys.* **6**, 29–40.
161. Cramer, H. E. (1959). Engineering estimates of atmospheric dispersal capacity. *Am. Ind. Hyg. Assoc. J.* **20**, 183–189.
162. Taylor, G. I. (1921). Diffusion by continuous movements. *Proc. London Math. Soc.* [2] **20**, 196–212.
163. Jones, J. I. P., and Pasquill, F. (1959). An experimental system for directly recording statistics of the intensity of atmospheric turbulence. *Quart. J. Roy. Meteorol. Soc.* **85**, 225–236.
164. Brock, F. V. and Provine, D. J. (1962). A standard deviation computer. *J. Appl. Meteorol.* **1**, 81–90.
165. Holzworth, G. C. (1964). Estimates of mean maximum mixing depths in the contiguous United States. *Monthly Weather Rev.* **92**, 235–242.
166. Spurr, G. (1959). The penetration of atmospheric inversions by hot plumes. *J. Meteorol.* **16**, 30–37.

167. Moses, H., and Strom, G. H. (1961). A comparison of observed plume rises with values obtained from well-known formulae. *J. Air Poll. Control Assoc.* **11**, 455–466.

168. Lucas, D. H., Moore, D. J., and Spurr, G. (1963). The rise of hot plumes from chimneys. *Int. J. Air Water Poll.* **7**, 473–500.

169. Rauch, H. (1964). Zur Schornstein-überhöhung. *Beitr. Phys. Atmosphäre* **37**, 132–158.

170. Munn, R. E., and Cole, A. F. W. (1965). Turbulence and diffusion in the wake of a building. *Preprint, 58th Ann. Meeting Air Poll. Control Assoc., Toronto, 1965.*

171. Halitsky, J. (1963). Gas diffusion near buildings. Rep No. 63–3, 119pp. Dept. Meteorol. and Oceanog., New York Univ., New York.

172. Blackadar, A. K. (1957). Boundary layer wind maxima and their significance for the growth of nocturnal inversions. *Bull. Am. Meteorol. Soc.* **38**, 283–290.

173. Hewson, E. W. (1955). Stack heights required to minimize ground concentrations. *Transact. Am. Soc. Mech. Engrs.* **77**, 1163–1172.

174. Monteith, J. L. (1957). Dew. *Quart. J. Roy. Meteorol. Soc.* **83**, 322–341.

175. Glaser, A. H., Elliott, W. P., and Druce, A. J. (1957). The study of small scale modification of air passing over inhomogeneous surfaces. Final Rep. No. AF 19(604)–997, 36pp. Res. Found., Texas A. & M., College Station, Texas.

176. Kazansky, A. B. (1960). Heat balance of open ice surface of Fedchenko glacier. *Bull. Acad. Sci. U. S. S. R., Geophys. Ser.* **12**, pp. 1883–1886. (English transl. from the Russian by Am. Geophys. Un.)

177. Liljequist, G. H. (1954). Radiation, and wind and temperature profiles over an antarctic snow field—a preliminary note *Proc. Meteorol. Conf., Toronto 1953*, pp. 78–87. Roy. Meteorol. Soc., London.

178. Lyubomirova, K. S. (1962). Certain features of the attenuation of solar radiation in an ice layer. *Bull. Acad. Sci., U. S. S. R., Geophys. Ser.* **5**, pp. 693–699. (English transl. from the Russian by Am. Geophys. Un.)

179. Takahashi, Y. (1960). On the puddles of Lützow-Holm Bay. *In* "Antarctic Meteorology," pp. 321–332. Pergamon Press, Oxford.

180. Murcray, W. B., and Echols, C. (1960). Some observations on the flow of heat through cold snow. *J. Meteorol.* **17**, 563–566.

181. Yen, Y.-C. (1962). Effective thermal conductivity of ventilated snow. *J. Geophys. Res.* **67**, 1091–1098.

182. Yen, Y.-C. (1963). Heat transfer by vapor transfer in ventilated snow. *J. Geophys. Res.* **68**, 1093–1101.

183. Scott, J. T., and Ragotzkie, R. A. (1961). Heat budget of an ice covered inland lake. Tech. Rep. No. 6, 41pp. Dept. Meteorol., Univ. of Wisconsin, Madison, Wisconsin.

184. Liljequist, G. H. (1958). Long-wave radiation and turbulent heat transfer in the antarctic winter and the development of surface inversions. *In* "Polar Atmospheric Symposium," (R. C. Sutcliffe, ed.), Part 1, pp. 167–181. Pergamon Press, Oxford.

185. Heigel, K. (1963). Die Reifbildung und die schneenahen Temperatur-Feuchte-Verhältnisse. *Meteorol. Runds.* **16**, 46–49.

186. Uchida, E. (1960). Diffusion des Wasserdampfs auf der Oberflächenschicht der Schneedecke. *Pap. Meteorol. and Geophys.* (*Meteorol. Res. Inst., Tokyo*) **11**, 196–211.

187. Orvig, S. (1963). Energy exchange at glacier surfaces. *In* Symposium on the heat exchange at snow and ice surfaces. Tech. Mem. No. 78, pp. 1–9. Natl. Res. Council, Ottawa.

188. LaChapelle, E. R. (1960). The Blue Glacier project. Rept. No. 477(18)(NR307–244), 54pp. Dept. Meteorol. and Climatol., Univ. of Washington, Seattle, Washington.

189. Gates, D. M. (1963). Leaf temperature and energy exchange. *Arch. Meteorol. Geophys., Biokl., B* **12**, 321–336.

190. Inoue, E. (1963). The environment of plant surfaces. *In* "Environmental Control of Plant Growth" (L. T. Evans, ed.), pp. 23–32. Academic Press, New York.

191. Rider, N. E. (1957). Water losses from various land surfaces. *Quart. J. Roy. Meteorol. Soc.* **83**, 181–193.

192. Rider, N. E. (1958). Discussion. *Quart. J. Roy. Meteorol. Soc.* **84**, 190.

193. Monteith, J. L. (1959). The reflection of short-wave radiation by vegetation. *Quart. J. Roy. Meteorol. Soc.* **85**, 386–392.

194. Graham, W. G., and King, K. M. (1961). Short-wave reflection coefficient for a field of maize. *Quart. J. Roy. Meteorol. Soc.* **87**, 425–428.

195. Allen, L. H., Yocum, C. S., and Lemon, E. R. (1962). The energy budget at the earth's surface. Interim Rep. No. 62–4, 23pp. N.Y.State Coll. Agr., Cornell Univ., Ithaca, New York. (3A99–27–055–08).

196. Uchijima, Z. (1962). Studies on the microclimate within the plant communities (1) on the turbulent transfer coefficient within plant layer. *J. Agr. Meteorol. Japan* **18**, 1–9 (in Japanese).

197. Waterhouse, F. L. (1955). Microclimatological profiles in grass cover in relation to biological problems. *Quart. J. Roy. Meteorol. Soc.* **81**, 63–71.

198. Penman, H. L., and Long, I. F. (1960). Weather in wheat: an essay in micrometeorology. *Quart. J. Roy. Meteorol. Soc.* **86**, 16–50.

199. Inoue, E. (1963). On the turbulent structure of airflow within crop canopies. *J. Meteorol. Soc. Japan* **41**, 317–325.

200. Rider, N. E. (1954). Evaporation from an oat field. *Quart. J. Roy. Meteorol. Soc.* **80**, 198–211.

201. Wright, J. L., and Lemon, E. R. (1962). The energy budget at the Earth's surface. Interim Rep. No. 62–7, 36pp. N.Y.State Coll. Agr., Cornell Univ. Ithaca, New York (3A99–27–005–08).

202. Uchijima, Z., and Wright, J. L. (1963). The energy budget at the Earth's surface. Interim Rep. No. 63–1, 50pp. N.Y.State Coll. Agr., Cornell Univ., Ithaca, New York (IAO–11001–B–021–08).

203. Barry, P. J., and Chamberlain, A. C. (1963). Deposition of iodine onto plant leaves from air. *Health Phys.* **9**, 1149–1157.

204. Barry, P. J. (1964). Some recent experiments on transfer of radioactive tracers from air to natural surfaces. Rep. No. CRL–86, AECL–2045, 11pp. Chalk River, Ontario.

205. Monteith, J. L. (1962). Measurement and interpretation of carbon dioxide fluxes in the field. *Neth. J. Agr. Sci.* **10**, 334–346.

206. Glesinger, E., ed. (1962). "Forest Influences," 307pp. F.A.O., Rome.

207. Heckert, L. (1959). Climatic conditions in deciduous forests. *Z. Meteorol.* **13**, 211–223.

208. Rauner, Yu. L. (1961). On the heat budget of a deciduous forest in winter. *Bull. Acad. Sci. U. S. S. R., Geog.* **4**, 83–90. (English transl. from the Russian, Meteorol. Branch, Toronto.)

209. Baumgartner, A. (1956). Untersuchungen über den Wärme- und Wasserhaushalt eines jungen Waldes. *Ber. Deut. Wetterd.* **5**, No. 28. 53pp.
210. Denmead, O. T. (1964). Evaporation sources and apparent diffusivities in a forest canopy. *J. Appl. Meteorol.* **3**, 383–389.
211. Waggoner, P. E., Pack, A. B., and Reifsnyder, W. E. (1959). The climate of shade. *Connecticut Agr. Expt. Sta., Bull.* **626**, 39pp.
212. Ziemer, R. R. (1964). Summer evapotranspiration trends as related to time after logging of forests in Sierra Nevada. *J. Geophys. Res.* **69**, 615–620.
213. Bayton, H. W. (1963). The penetration and diffusion of a fine aerosol in a tropical rain forest. Ph.D. Thesis, pp. 118–209. Univ. of Michigan, Ann Arbor, Michigan.
214. Skau, C. M., and Swanson, R. H. (1963). An improved heat pulse velocity meter as an indicator of sap speed and transpiration. *J. Geophys. Res.* **68**, 4743–4749.
215. Rauner, Yu. L. (1958). Some results of heat budget measurements in a deciduous forest. *Bull. Acad. Sci. U. S. S. R., Geog.* **5**, 79–86. (English transl. from the Russian, Meteorol. Branch, Toronto.)
216. Philip, J. R. (1964). Sources and transfer processes in the air layers occupied by vegetation. *J. Appl. Meteorol.* **3**, 390–395.
217. Ovey, C. D. (1959). Some effects of weather and climate on surface life in the open oceans. *Weather* **14**, 339–344.
218. Beeton, A. M. (1962). Variations in radiant energy and related ocean temperatures. *Proc. Fifth Conf. on Great Lakes Res. 1962*, pp. 68–78. Univ. of Michigan, Great Lakes Res. Div., Ann Arbor, Michigan.
219. Olson, B. E. (1962). Variations in radiant energy and related ocean temperatures. *J. Geophys. Res.* **67**, 4705–4711.
220. Klevantsova, V. A., Bortkovskii, R. S., and Preobrazhenskii, L. Yu. (1964). Methods of gradient measurements at sea. *Trans. Main. Geophys. Obs., U. S. S. R.,* **150**, 85–94. (English transl. from the Russian, Meteorol. Branch, Toronto.)
221. Deacon, E. L., Sheppard, P. A., and Webb, E. K. (1956). Wind profiles over the sea and the drag at the sea surface. *Australian J. Phys.* **9**, 511–541.
222. Hamilton, H. L. Jr. (1962). Spatially continuous measurements of temperature profiles through an air-water interface. Final Rep. No. DA–36–039–SC–80282, pp. 47–68. Dept. Meteorol., Univ. of Wisconsin, Madison, Wisconsin.
223. Gates, D. M., Vetter, M. J., and Thompson, M. C. (1963). Measurement of moisture boundary layers and leaf transpiration with a microwave refractometer. *Nature* **197**, 1070–1072.
224. McAlister, E. D. (1963). Temperature gradients in the superficial layer of the ocean. Abst. *Proc. 13th General Assembly, IUGG-IAMAP, Berkeley, Calif., 1963,* p. 125.
225. Hasse, L. (1963). On the cooling of the sea surface by evaporation and heat exchange. *Tellus* **15**, 363–366.
226. Kondo, J. (1962). Evaporation from extensive surfaces of water. *Sci. Rep. Tohoku Univ. Geophys.* **14**, 107–119.
227. Deacon, E. L., and Webb, E. K. (1962). Small-scale interactions. *In* "The Sea," (M. N. Hill, ed.) Vol. 1, pp. 43–87. Wiley (Interscience), New York.
228. Fitzgerald, L. M. (1963). Wind-induced stresses on water surfaces: A wind-tunnel study. *Australian J. Phys.* **16**, 475–489.

229. Vinogradova, O. P. (1959). Tangential wind stress above a disturbed sea surface. *Bull. Acad. Sci. U. S. S. R., Geophys. Ser.* **11**, pp. 1646–1655. (English transl. from the Russian by Am. Geophys. Un.)

230. Stewart, R. W. (1961). The wave drag of wind over water. *J. Fl. Mech.* **10**, 189–194.

231. Dutton, J. A., and Bryson, R. A. (1962). Heat flux in Lake Mendota. *Limnolog. and Oceanog.* **7**, 80–97.

232. Colón, J. A. (1963). Seasonal variations in heat flux from the sea surface to the atmosphere over the Caribbean Sea. *J. Geophys. Res.* **68**, 1421–1430.

233. Bruce, J. P., and Rodgers, G. K. (1962). Water balance of the Great Lakes system. *In* Great Lakes basin (Abst.). Publ. No. 71, pp. 41–69. Am. Assoc. Advance. Sci. Washington, D.C.

234. Swinbank, W. C. (1959). Evaporation from oceans. Sci. Rep. No. 12, AF19(604)–2179, 15pp. Dept. Meteorol., Univ. of Chicago, Chicago, Illinois.

235. Palmén, E. (1963). Computation of the evaporation over the Baltic Sea from the flux of water vapor in the atmosphere. *Proc. 13th General Assembly IASH-IUGG (Commission for Evaporation). Berkeley, Calif., 1963.* Publ. No. 62, pp. 244–252.

236. Ninomiya, K. (1964). Heat budget over the Japan Islands during the period of heavy snow storm. *Pap. Meteorol. and Geophys. (Meteorol. Res. Inst., Tokyo)* **15**, 52–69.

237. Pandolfo, J. (1961). Power spectrum analyses of turbulent surface winds over water under inversion conditions. Rep. No. 285–(03), 67pp. Dept. Meteorol. and Oceanog., New York Univ., New York.

238. Brocks, K., and Hasse, L. (1963). Fluctuation measurements made at sea with a gyroscopic stabilised floating mast. Abst. in *Proc. 13th General Assembly IUGG-IAMAP Berkeley, Calif.,* 1963, p. 117.

239. Slade, D. H. (1962). Atmospheric dispersion over Chesapeake Bay. *Monthly Weather Rev.* **90**, 217–224.

240. Pearce, R. P. (1956). Discussion on the calculation of a sea-breeze circulation in terms of the differential heating across the coast line. *Quart. J. Roy. Meteorol. Soc.* **82**, 239.

241. Wallington, C. E. (1961). "Meteorology for Glider Pilots," 284pp. John Murray, London.

242. Wexler, R. (1946). Theory and observations of land and sea breezes. *Bull. Am. Meteorol. Soc.* **27**, 272–287.

243. Dexter, R. V. (1954). Some local temperature variations in Halifax, N.S. Rep. No. Tec-184, 11pp. Meteorol. Branch, Toronto (CIR-2473).

244. Estoque, M. A. (1961). A theoretical investigation of the sea breeze. *Quart. J. Roy. Meteorol. Soc.* **87**, 136–146.

245. Estoque, M. A. (1962). The sea breeze as a function of the prevailing synoptic situation. *J. At. Sci.* **19**, 244–250.

246. Fisher, E. L. (1960). An observational study of the sea breeze, *J. Meteorol.* **17**, 645–660,

247. Frizzola, J. A., and Fisher, E. L. (1963). A series of sea breeze observations in the New York City area. *J. Appl. Meteorol.* **2**, 722–739.

248. Hatcher, R. W., and Sawyer, J. S. (1947). Sea breeze structure with particular reference to temperature and water vapor gradients and associated radio ducts. *Quart. J. Roy. Meteorol. Soc.* **73**, 391–406.

249. Craig, R. A., Katz, I., and Harney, P. J. (1945). Sea-breeze cross sections from psychrometric measurements. *Bull. Am. Meteorol. Soc.* **26**, 405–411.
250. Stern, M. E., and Malkus, J. S. (1953). The flow of a stable atmosphere over a heated island, Part 2. *J. Meteorol.* **10**, 105–120.
251. Scorer, R. S. (1959). The behaviour of chimney plumes. *Int. J. Air Water Poll.* **1**, 198–220.
252. Klassen, W. (1962). Micrometeorological observations in the North Saskatchewan river valley at Edmonton. Rep. No. Tec-408, 24pp. Meteorol. Branch, Toronto (CIR-3652).
253. Buettner, K. J. K., and Thyer, N. (1962). Valley winds in Mt. Rainier National Park. *Weatherwise* **15**, 63–68.
254. Wilkins, E. M. (1955). A discontinuity surface produced by topographic winds over the upper Snake River Plain, Idaho. *Bull. Am. Meteorol. Soc.* **36**, 397–408.
255. Koch, H. G. (1961). Die warme Hangzone. *Z. Meteorol.* **15**, 151–171.
256. Davidson, B. (1961). Valley wind phenomena and air pollution problems. *J. Air Poll. Control Assoc.* **11**, 364–368.
257. Davidson, B. and Rao, P. K. (1963). Experimental studies of the valley-plain wind. *Int. J. Air Water Poll.* **7**, 907–923.
258. Ayer, H. S. (1961). On the dissipation of drainage wind systems in valleys in morning hours. *J. Meteorol.* **18**, 560–563.
259. McCormick, R. A. (ed.) (1961). Symposium: air over cities. Tech. Rep. No. A62–5, 290pp. U.S. Public Health Service, Cincinnati.
260. Kratzer, P. A. (1956). Das Stadtklima. *Wissenschaft (Braunschweig)* **90**, 184pp.
261. Landsberg, H. E. (1961). City air—better or worse? *In* Symposium: air over cities. Tech. Rep. A62–5, pp. 1–22. U.S. Public Health Service, Cincinnati.
262. Sheppard, P. A. (1958). The effect of pollution on radiation in the atmosphere. *Int. J. Air Water Poll.* **1**, 31–43.
263. Hand, I. F. (1943). Transmission of the total and the infrared component of solar radiation through a smoky atmosphere. *Bull. Am. Meteorol. Soc.* **24**, 201–204.
264. Mateer, C. L. (1961). Note on the effect of the weekly cycle of air pollution on solar radiation at Toronto. *Int. J. Air Water Poll.* **4**, 52–54.
265. Meetham, A. R. (1945). Atmospheric pollution in Leicester. DSIR Tech. Paper No. 1, 161pp. H.M. Stationery Office, London.
266. Blackwell, M. J., Eldridge, R. H., and Robinson, G. D. (1954). Estimation of the reflection and absorption of solar radiation by a cloudless atmosphere from recordings at the ground with results from Kew Observatory. Rep. No. MRP–894, 11pp. Air Ministry, London.
267. Roach, W. T. (1961). Some aircraft observations of fluxes of solar radiation in the atmosphere. *Quart. J. Roy. Meteorol. Soc.* **87**, 346–363.
268. Bryson, R. A. (1963). Personal communication.
269. McCormick, R. A., and Baulch, D. M. (1962). The variation with height of the dust loading over a city as determined from the atmospheric turbidity. *J. Air Poll. Control Assoc.* **12**, 492–496.
270. Chandler, T. J. (1962). London's urban climate. *Geog. J.* **127**, 279–302.
271. Gold, L. (1954). Discussion on physiological effects of smog. *Quart. J. Roy. Meteorol. Soc.* **80**, 273–274.
272. Duckworth, F. S., and Sandberg, J. S. (1954). The effect of cities upon horizontal and vertical temperature gradients. *Bull. Am. Meteorol. Soc.* **35**, 198–207.

273. DeMarrais, G. A. (1961). Vertical temperature difference observed over an urban area. *Bull. Am. Meteorol. Soc.* **42**, 548–554.
274. Munn, R. E., Emslie, J. H., and Wilson, H. J. (1963). A preliminary analysis of the inversion climatology of southern Ontario. Rep. No. Tec-466, 13pp. Meteorol. Branch, Toronto (CIR-3834).
275. Mitchell, J. M. (1961). The thermal climate of cities. *In* Symposium: air over cities. Tech. Rep. No. A62-5, pp. 131–145. U.S. Public Health Service, Cincinnati.
276. Robertson, G. W. (1955). Low-temperature fog at the Edmonton Airport as influenced by moisture from the combustion of natural gas. *Quart. J. Roy. Meteorol. Soc.* **81**, 190–197.
277. Pooler, F. (1963). Airflow over a city in terrain of moderate relief. *J. Appl. Meteorol.* **2**, 446–456.
278. Davenport, A. G. (1963). The relationship of wind structure to wind loading. Preprint, *Int. Conf. on Wind Effects on Buildings and Structures, 1963*, 51pp. Natl. Phys. Lab., Teddington, England.
279. Frederick, R. H. (1961). A study of the effect of tree leaves on wind movement. *Monthly Weather Rev.* **89**, 39–44.
280. Wainwright, C. W. K., and Wilson, M. J. G. (1962). Atmospheric pollution in a London park. *Int. J. Air Water Poll.* **6**, 337–347.
281. Whiten, A. J. (1956). The ventilation of Oxford Circus. *Weather* **11**, 227–229.
282. Lucas, D. H. (1958). The atmospheric pollution of cities. *Int. J. Air Water Poll.* **1**, 71–86.
283. Turner, D. B. (1964). A diffusion model for an urban area. *J. Appl. Meteorol.* **3**, 83–91.
284. Rossano, A. T., and Schell, N. E. (1957). Procedures for making an inventory of air pollution emissions. *Proc. Golden Jubilee Meeting Air Poll. Control Assoc., Pittsburgh, 1957*, pp. 11–17.
285. Ministry of Housing and Local Development (1963). "Clean Air Act 1956; Memorandum on Chimney Heights," 5pp. H.M. Stationery Office, London.
286. Blanc, M. L., Geslin, H., Holzberg, I. A., and Mason, B. (1963). Protection against frost damage. Tech. Note No. 51, 62pp. World Meteorol. Org., Geneva.
287. van Eimern, E. (1964). Windbreaks and shelterbelts. Tech. Note. No. 70, 115pp. World Meteorol. Org., Geneva.
288. Ramdas, L. A. (1957). Natural and artificial modification of microclimate. *Weather* **12**, 237–240.
289. Williams, G. P., and Gold, L. W. (1963). The use of dust to advance the break-up of ice on lakes and rivers. *In* Proc. Ann. meeting eastern snow conf. Tech. Paper No. 165, pp. 31–56. Natl. Research Council, (Div. Building Res.), Ottawa.
290. Brooks, F. A., and Schultz, H. B. (1958). Observation and interpretation of nocturnal density currents. *Proc. Symp. on Climatol. and Microclimatol., 1958, Canberra*, pp. 272–277. UNESCO, Paris.
291. Belyaev, V. I., and Pavlova, I. S. (1962). On the possibility of influencing the weather by artificial dissipation of cloudiness. *Bull. Acad. Sci. U. S. S. R., Geophys. Ser.* **1**, pp. 129–133. (English transl. from the Russian by Am. Geophys. Un.)
292. Jarvis, N. L. (1963). The effect of monomolecular films on the surface temperature and convection motion. Abst. in *Proc. 13th General Assembly IUGG-IAMAP, Berkeley, Calif., 1963*, p. 125.

293. Schultz, H. B. (1962). The interaction of the macro- and microclimatic factors contributing to the success of wind machines for frost protection in southern California. *In* "Biometeorology" (S.W. Tromp, ed.), pp. 614–629. Pergamon Press, Oxford.

294. Angus, D. E. (1962). Frost protection experiments using wind machines. Tech. Paper No. 12, 44pp. C.S.I.R.O. Meteorol. Phys. Australia.

295. Smal'ko, Ya. A. (1955). The zones of wind speed reduction by forest belts of different types. *Bull. Acad. Sci. U. S. S. R., Geog.* **5**, 44–49. (English transl. from the Russian, Meteorol. Branch, Toronto.)

296. Konstantinov, A. R., and Vorontsov, P. A. (1961). The effect of forest belts on wind and eddy exchange in the atmosphere. *Ukr. Sci./Res. Hydrometeorol. Inst.* **26**, 99–110. (English transl. from the Russian, Meteorol. Branch, Toronto.)

297. Jackson, A. A. (1959). The ecoclimatology of Dutch lights. *Weather* **14**, 117–123, 155–162.

Author Index

Numbers in parentheses are reference numbers and indicate that an author's work is referred to although his name is not cited in the text. Numbers in italic show the page on which the complete reference is listed.

Allen, C. W., 9 (2), *221*
Allen, L. H., 146, *230*
Angell, J. K., 79 (103), *225*
Angus, D. E., 32 (39, 41), 212 (294), *222,* *235*
Ayer, H. S., 193, *233*

Ball, F. K., 73 (94), 74 (94), *225*
Barry, P. J., 152, 153, *230*
Bauer, K. G., 16 (12), *221*
Baulch, D. M., 197, *233*
Baumgartner, A., 156, 157 (209), 162, 163, 165, *231*
Bayton, H. W., 159 (213), 160 (213), 161 (213), *231*
Beeton, A. M., 170 (218), 171 (218), *231*
Belyaev, V. I., 211, *234*
Blackadar, A. K., 64 (73), 65 (73), 127 (172), *224,* *229*
Blackwell, M. J., 196, 197 (266), *233*
Blanc, M. L., 32 (42), 208 (286), *223,* *234*
Bonner, J., 30 (35), 31 (35), *222*
Bortkovskii, R. S., 171 (220), 174 (220), *231*
Bowen, I. S., 5 (1), *221*
Bradley, E. F., 114 (156), 131 (156), *228*
Brier, G. W., 77 (97), *225*
Brock, F. V., 113, 123 (164), *228*
Brocks, K., 178 (238), *232*
Brooks, F. A., 60 (64), 61 (64), 108, 111 (144), 112 (144), 210, *224,* *227,* *234*
Bruce, J. P., 177, *232*
Brunt, D., 58, *223*
Bryson, R. A., 176, 197 (268), 198 (268), *232,* *233*

Budyko, M. I., 16, 28, 130 (9), 208 (9), *221*
Buettner, K. J. K., 27 (29), 192, *222,* *233*
Businger, J. A., 98 (137), *227*

Calder, K. L., 58, *223*
Carson, J. E., 37 (49), 39 (49), *223*
Chamberlain, A. C., 152, *230*
Chandler, T. J., 199, 200 (270), 202 (270), 204, 207, *233*
Clayton, W. H., 94 (133), 95 (133), *227*
Cole, A. F. W., 126 (170), *229*
Colón, J. A., 177, *232*
Craig, R. A., 187 (249), *233*
Cramer, H. E., 69 (79), 75 (96), 121, *224,* *225,* *228*
Crawford, T. V., 95, *227*

Dalrymple, P. C., 58 (60), 65 (60), 138, 139 (60), 140 (60), *223*
Daubek, H. G., 103 (141), 104, *227*
Davenport, A. G., 71 (85), 204, *225,* *234*
Davidson, B., 24 (26), 25 (26), 26 (26), 29 (26), 30 (26), 34 (26), 37 (26), 58 (26), 65 (26), 80 (26), 90 (26), 94 (26), 143, 193 (256, 257), 196 (26), *222,* *232*
Deacon, E. L., 61, 172 (221), 175, *224,* *231*
De Marrais, G. A., 64, 203 (273), *224,* *234*
Denmead, O. T., 157 (210), 165, *231*
Dexter, R. V., 182, 183 (243), *232*
Druce, A. J., 131 (175), *229*
Duckworth, F. S., 200 (272), 201 (272), 202 (272), 203 (272), *233*
Dutton, J. A., 16 (12), 176, *221,* *232*

237

Dyer, A. J., 88 (121, 124), 96 (121, 124), 98 (121), 114 (154, 157), 115 (157), 116, *226, 228*

Echols, C., 136, *229*
Ehlig, C. F., 32 (37), *222*
Elagina, L. G., 98, *227*
Eldridge, R. H., 196 (266), 197 (266), *233*
Elliott, W. P., 90, 108, 131 (175), *227, 228, 229*
Ellison, T. H., 83, 86 (109), 87, 94, *226*
Emslie, J. H., 203 (274), *234*
Estoque, M. A., 116, 184 (244, 245), 185 (245), 186 (245), *228, 232*

Fisher, E. L., 184, *232*
Fitzgerald, L. M., 175 (228), *232*
Franssila, M., 21 (22), *222*
Frederick, R. H., 205, *234*
Frenkiel, F. N., 69, *224*
Fritz, S., 13 (7), 16 (7), *221*
Frizzola, J. A., 184, *232*
Funk, J. P., 20, 21, 48 (52), 81, *221, 222, 223*

Gaevskaya, G. N., 90, *227*
Gardner, W. R., 32 (37), *222*
Gates, D. M., 144, 172, *230, 231*
Geiger, R., 18 (13), 134 (13), 146, 155, 190, 195, *221*
Geslin, H., 208 (286), *234*
Gifford, F., 72, *225*
Gill, G. C., 80 (106), 106, *226, 227*
Glaser, A. H., 131 (175), *229*
Glesinger, E., 155 (206), 156 (206) 158 (206), *230*
Goddard, W. B., 65 (75), *224*
Godson, W. L., 9, *221*
Gold, L., 199, *233*
Gold, L. W., 210, *234*
Graham, W. G., 145 (194), *230*
Grant, H. L., 72, *225*
Gurvich, A. S., 65 (76), 88 (123), *224, 226*

Halitsky, J., 101, 126, *227, 229*
Hamilton, H. L., Jr., 21, 172 (222), 173 (222), *222, 231*

Hand, I. F., 196, *233*
Harney, P. J., 187 (249), *233*
Harrell, W., 23 (24), *222*
Hasse, L., 174, 178 (238), *231, 232*
Hatcher, R. W., 187, *232*
Hay, J. S., 78 (102), *225*
Heckert, L., 155 (207), 156 (207), 159, *230*
Heigel, K., 138 (185), *229*
Hewson, E. W., 47 (51), 127 (51), 128, *223, 229*
Hickman, K. C., 92, *227*
Hinze, J. O., 70 (84), 79, 101, *225*
Holzberg, I. A., 208 (286), *234*
Holzman, B., 93, *227*
Holzworth, G. C., 124 (165), *228*

Imai, K., 108 (145), *227*
Inoue, E., 108, 144, 148, 149, 157, *227, 230*
Islitzer, N. F., 21 (22), *222*
Isobe, S., 108 (145), *227*

Jackson, A. A., 215 (297), *235*
Jarvis, N. L., 212, *234*
Jen-Hu-Chang, 15 (8), *221*
Jex, H. R., 78 (101), 80 (107), *225, 226*
John, P. T., 27 (28), *222*
Jones, J. I. P., 123 (163), *228*

Kaganov, M. A., 37 (47), *223*
Kaimal, J. C., 98 (137), *227*
Katz, I., 187 (249), *233*
Kaulin, N. Ya., 45 (50), *223*
Kazansky, A. B., 87, 134 (176), 139, 142, *226, 229*
King, K. M., 32 (38), 145 (194), *222, 230*
Klassen, W., 192 (252), 193, *233*
Klevantsova, V. A., 171 (220), 174, *231*
Koch, H. G., 193, *233*
Kofoed-Hansen, O., 79, *225*
Kolmogorov, A. N., 69 (81), 71, *224, 225*
Kondo, J., 175, *231*
Kondrati'ev, K. Y., 90 (128), *227*
Konstantinov, A. R., 213 (296), *235*
Kraichnan, R. H. *80*
Kratzer, P. A., 195, 199, 202 (260), 203 204 (260), 207, *233*

Kraus, H., 19 (16), 90, 91 (16), *221*
Kutzbach, J. E., 109, 110 (149), *228*

La Chapelle, E. R., 142 (188), *230*
Landsberg, H. E., 196, *233*
Langhaar, H. L., 54 (55), 55 (55), *223*
Latimer, J. R., 16, 21, *221*, 222
Le Hau, H. H., 196 (26), *222, 223, 224, 228*
Lemon, E. R., 146 (195), 151 (201), 153, 166 (201), *230*
Lettau, H. H., 24 (26), 25 (26), 26 (26), 29 (26), 30 (26), 34 (26), 36, 37 (26, 44), 53, 58 (26, 57, 60), 61, 62, 65 (26, 60), 80 (26), 87, 90 (26), 94 (26), 112 (150), 138 (60), 139 (60), 140 (60), 143, 209 (150), *222*
Liljequist, G. H., 134, 138, 140, *229*
List, R. J., 9 (4), 10 (4), *221*
Long, I. F., 147 (198), 150, 151 (198), *230*
Lucas, D. H., 125, 207 (282), *229, 234*
Lumley, J. L., 79, *225*
Lyubomirova, K. S., 134, *229*

McAlister, E. D., 173, *231*
McCormick, R. A., 195 (259), 197, *233*
MacCready, P. B., 78 (101), 80 (107), *225, 226*
McIlroy, I. C., 27 (27), 28 (27), 29, 31 (27), 32 (27, 41), 131 (27), *222*
McVehil, G. E., 64 (73), 65 (73), *224*
Malkus, J. S., 188, *233*
Maruyama, E., 28 (32), 29 (32), 40 (32), *22*
Mason, B., 208 (286), *234*
Mateer, C. L., 9, 196, *221, 233*
Meetham, A. R., 196, *233*
Mistry, P. D., 94 (133), 95 (133), *227*
Mitchell, J. M., 203, *234*
Miyake, M., 114, *228*
Moilliet, A., 72 (88), *225*
Molga, M., 30 (33), 32 (33), *222*
Monin, A. S., 58, 62, 81, 86 (116), 87, 120, 138, *223, 226, 228*
Monteith, J. L., 24, 130 (174), 145 (193), 153 (205), 154 (205), *222, 229, 230*
Moon, P., 12 (5), *221*
Moore, D. J., 125 (168), *229*
Moses, H., 37 (49), 39 (49), 103 (141), 104, 125, *223, 227, 229*

Munn, R. E., 13 (6), 15 (6), 63 (71), 64 (71), 126 (170), 203 (274), *221, 224*, 229, *234*
Murcray, W. B., 136, *229*

Nicolet, M., 16 (10), *221*
Ninomiya, K., 177, *232*
Novikov, E. A., 69 (83), *225*

Obukhov, A. M., 58, 62, 69 (82), 73, 81, 138, *223, 224, 225*
Olson, B. E., 171 (219), *231*
Orvig, S., 141 (187), *229*
Ovey, C. D., 168, *231*

Pack, A. B., 157 (211), *231*
Padmanabhamurty, B., 37 (48), 38 (48), *223*
Palmén, E., 177, *232*
Pandolfo, J., 178 (237), *232*
Panofsky, H. A., 64 (73), 65, 68 (78), 72 (78), 75, 77 (97), 78 (99), 79, 87, 93 (99) 108, *224, 225, 226, 228*
Pasquill, F., 68 (78), 71, 72 (78, 86), 73, 78 (100, 102), 79 (86), 118, 123, (163) *224, 225, 228*
Pavlova, I. S., 211, *234*
Pearce, R. P., 180, *232*
Pelton, W. L., 32 (40), *222*
Penman, H. L., 30 (34), 31 (36), 147 (198), 150, 151 (198), *222, 230*
Perepelkina, A. V., 88 (122), *226*
Philip, J. R., 27 (30), 37 (46), 39, 113, 114 (156), 131 (153, 156), 165, *222, 223, 228*, 231
Pooler, F., 204, *234*
Portman, D. J., 37 (45), 113 (151), *223, 228*
Preobrazhenskii, L. Yu., 171 (220), 174 (220), *231*
Priestley, C. H. B., 63, 73 (93), 84, 85 (69), 108, 115, *224, 225, 226*
Provine, D. J., 123 (164), *228*
Pruitt, W. O., 32 (39), 96 (136), 97 (136), *222, 227, 228*

Ragotzkie, R. A., 137 (183), *229*
Ramdas, L. A., 208 (288), 209 (288), 211 (288), *234*

Rao, P. K., 193 (257), *233*
Rao., V. R. K., 75 (96), *225*
Rauch, H., 125, *229*
Rauner, Yu.L., 156 (208), 161, 163, 164 (208), *230*, *231*
Reifsnyder, W. E., 157 (211), *231*
Richards, T. L., 63 (71), 64 (71), *224*
Richardson, E. A., 23 (24), *222*
Richardson, L. F., 82, *226*
Rider, N. E., 32 (43), 61, 94, 105, 106 (142), 113, 114, 131 (153, 156), 143, 145 (191, 192), 149 (200), *223*, *224*, *227*, *228*, *230*
Roach, W. T., 197, *233*
Robertson, G. W., 204, *234*
Robinson, G. D., 19, 83, 196 (266), 197 (266), *221*, *226*, *233*
Rodgers, G. K., 177, *232*
Rodhe, B., 48 (53), *223*
Romanova, E. N., 45 (50), *223*
Rosenstock, Yu.L., 37 (47), *223*
Rossano, A. T., 207, *234*

Sandberg, J. S., 200 (272), 201 (272), 202 (272), 203 (272), *233*
Sarson, P. B., 28, *222*
Sawyer, J. S., 187, *232*
Sceicz, G., 24, *222*
Schell, N. E., 207 (284), *234*
Schultz, H. B., 210, 212 (293), *234*, *235*
Scorer, R. S., 190 (251), *233*
Scott, J. T., 137 (183), *229*
Sechrist, F., 92, *227*
Sellers, W. D., 87, *226*
Senderikhina, I. L., 83, 94, *226*
Shaw, J. H., 18 (14), *221*
Sheppard, P. A., 65 (74), 114, 172 (221), 174, 175, 196, *224*, *228*, *231*, *233*
Skau, C. M., 162 (214), *231*
Slade, D. H., 178 (239), *232*
Slatyer, R. O., 27 (27), 28 (27), 29, 31 (27), 32, 131 (27), *222*
Smal'ko, Ya.A., 213 (295), 214 (295), *235*
Spurr, G., 124 (166), 125 (168), *228*, *229*
Stearns, C. R., 112 (150), 209 (150), *228*
Stern, M. E., 188, *233*
Sternberg, J., 88, *226*

Stewart, R. W., 59 (63), 68, 69 (83), 72 (88), 86 (115), 176 (230), *224*, *225*, *226*, *232*
Strom, G. H., 125, *229*
Subrahmanyan, V. P., 37 (48), 38 (48), *223*
Suomi, V. E., 21, 32 (38), *222*
Sutton, O. G., 57, 59, 112, 115, 119, 122, *223*
Swanson, R. H., 162 (214), *231*
Swinbank, W. C., 78 (98), 83, 84, 87, 114, 143, 177, *225*, *226*, *232*

Takahashi, Y., 135 (179), *229*
Takeuchi, K., 73, *225*
Tani, N., 108 (145), *227*
Tanner, C. B., 32 (38), *222*
Taylor, G. I., 75, 76, 121, *225*, *228*
Taylor, R. J., 59, 63 (70), 86, 87, 88 (121), 89, 96 (121, 135), 99 (121), 108, *223*, *224*, *226*, *227*, *228*
Thompson, M. C., 172 (223), *231*
Thornthwaite, C. W., 93 *227*
Thyer, N., 192, *233*
Townsend, A. A., 84 (113), 85, 108, *226*, *228*
Truhlar, E. J., 13 (6), 15 (6), *221*
Turner, D. B., 207 (283), *234*

Uchida, E., 141, *229*
Uchijima, Z., 146, 149 (196), 150 (196), 151 (202), 166 (202), *230*

van Eimern, E., 208 (287), 210 (287), 213 (287), 215 (287), *234*
Vetter, M. J., 172 (223), *231*
Vinogradova, O. P., 175, 178 (229), 212 (229), *232*
Vorontsov, P. A., 213 (296), *235*

Waggoner, P. E., 157, *231*
Wainwright, C. W. K., 206 (280), *234*
Wallington, C. E., 181 (241), *232*
Wandel, C. F., 79, *225*
Waterhouse, F. L., 147 (197), 148 (197), *230*
Webb, E. K., 172 (221), 175, *231*
Wexler, R., 182 (242), *232*

Whiten, A. J., 207, *234*
Wilkins, E. M., 193 (254), *233*
Williams, G. P., 210, *234*
Wilson, H. J., 203 (274), *234*
Wilson, M. J. G., 206 (280), *234*
Wollaston, S. H., 58 (60), 65 (60), 138 (60), 139 (60), 140 (60), *223*
Woodruff, N. P., 99, *227*
Wright, J. L., 151, 153, 166 (201, 202), *230*

Yaglom, A. M., 73, *225*
Yakushevskaya, K. E., 90 (128), *227*
Yamamoto, G., 18, 87, 208 (13), *221*, *226*
Yen, Y.-C., 136, *229*
Yocum, C. S., 146 (195), *230*

Ziemer, R. R., 158 (212), *231*
Zingg, A. W., 99, *227*
Zubkovskii, S. L., 72, *225*
Zwang, L. R., 88 (123), 89 (126), 98, *226*

Subject Index

A

Ablation, 141
Adiabatic, *see* Lapse rate
Advection, 42, 47, 112–115, 172, 176, 178
Air pollution, *see* Atmospheric pollution
Air temperature, 42–52, 85, 130, 144, 158–159, 171, 181–186, 194, 199–204, 209, 211, *see also* Lapse rate
measurement of, 49, 52, 98
Albedo, 14–15, *see also* Radiation
Anemometer, *see* Turbulence, measurement of, Wind, measurement of
Apparent surface radiative temperature, *see* Surface temperature
Atmospheric pollution, 47, 118–128, 194, 195–207
Austausch coefficient, 36, 90–91
Autocorrelations, *see* Correlation coefficients
Azimuth angle, *see* Bidirectional vane

B

Beer's law, 13, 134, 146, 157, 169–170, 197–198
Bidirectional vane, 66–67, 79–80, 123, 126–127
Blackbody, 8, *see also* Radiation
Bowen ratio, 5, 94–95, 113

C

Carbon dioxide, 29, 31, 92, 97, 153–154
City climate, 195–207
Coherence, 77
Conduction, *see* Heat transfer
Conductivity, 33–34

Convection,

forced, 42, 84, 95
free, 43, 84–85, 95
natural, 42, 85
penetrative, 85
Coriolis force, 1, 180, 182, 185, 189
Correlation coefficients, 74–76, 78, 121–122
Cospectrum, 77
Covariances, 68, 77–78, 88, 96

D

Deacon numbers, 61–62, 139–140
Density
of air, 43, 77–78, 88–89, 96
of soils, 34–36
of water, 167
Dew, 130–131
Diffusion, *see* Atmospheric pollution
Diffusivity
momentum, 57, 82–84, 94–95, 176
soil heat, 36
soil moisture, 39–40
turbulent heat, 82–84, 94–95, 132, 149, 176
water vapor, 93–95, 132, 149, 174, 176
Dimensional analysis, 54–56, 59, 62, 71, 81, 85
Dissipation of energy, *see* Viscous dissipation
Distance constant, 80
Divergence, *see* Flux divergence
Downwash, 125–127, 180
Drag coefficient, 56, 175

E

Effective stack height, 124–125

Elevation angle, *see* Bidirectional vane
Emissivity, 17, *see also* Radiation
Energy balance, 2–5, 37, 129–132, 136–137, 144–145, 149–151, 163–166, 177
Ergodic principle, 122
Eulerian-Lagrangian comparisons, 78–79, 123
Evaporation, 3, 92–98, 141, 177, 211–212
Evapotranspiration, *see* Transpiration
Evapotron, 88, 96
Extinction coefficient, 13, *see also* Beer's law

F

Field capacity, 27–28
Flux, 3
Flux divergence
    of heat, 19–21, 62, 89–91, 139, 146, 149–150, 165–166
    of momentum, 58, 62, 139
    of water vapor, 49, 93
Fog, 48–49, 204, 211
Forced convection, *see* Convection
Forest clearings, *see* Shelter belts
Forests, 155–166
Free convection, *see* Convection
Freezing, 137, 158, 168
Friction velocity, 57
Frost penetration, *see* Freezing
Frozen turbulence, 68–69, 75, 89
Fumigation, 47, 127–128, 194

G

Geostrophic wind, 6, 53, 179–180, 183, 187, 189
Glaciers, 140, 142, 191
Graybody, 17
Greenhouse effect, 18, 135, 215–216

H

Heat island of cities, *see* Urban heat island
Heat storage, 135–137, 162–163, 176, 199
Heat transfer
    in air, *see* Convection, Flux divergence of heat, Latent heat, Radiation
    in soil, 3, 33–37, 158
    in snow and ice, 135–137

Homogeneity, 68
Humidity, 49, 141–142, 147–148, 161–162, 172–174, 187, 204, 211, *see also* Latent heat
    measurement of, 49, 52, 98
    mixing ratio, 49–50
    specific humidity, 49
    vapor pressure, 49

I

Ice, 109–112, 133–142, 209–210
Inertial subrange, 72, 73, 76
Intermittency, 66–67, 69, 100, 181, 192–193
Inversion, *see* Lapse rate
Isotropy, 68
    local isotropy, 68, 70–73

K, L

Kármán's constant, *see* Von Kármán's constant
Lagrangian, 78–79, 122–123
Lake, 109–110, 137, 167–188
Laminar sublayer, 88, 151–153
Land and sea breezes, 179–188
Langley, 3
Lapse rate,
    adiabatic, 43–45, 114, 138, 171
    inversion, 43–48, 85, 89, 113, 130, 138, 147, 158–159 172, 180–181, 194, 203, 209, 212–213
    in water, 167, 172
    superadiabatic, 43–45, 85, 114, 138, 158–159, 180, 213
Latent heat, 3–5, 93–97, 112–116, 131–132, 136–137, 141–142, 144, 148–151, 163–166, 177
Leaf area index, 145, 146, 213
Local isotropy, *see* Isotropy
Long-wave radiation, *see* Radiation
Lysimeter, 32, 95, 97

M

Mass transfer, *see* Atmospheric pollution, Evaporation, Latent heat
Mixing ratio, *see* Humidity
Modification of local weather, 208–216

Monin-Obukhov length, 62–64, 81–82, 84–87
Monomolecular films, 211–212

N, O

Natural convection, *see* Convection
Net radiation, *see* Radiation
Oceans, 167–188
Optical air mass, 12–13, 18, *see also* Beer's law
Optical density, 13, *see also* Beer's law
Ozone, 10, 97, 127

P

Percolation, 28, 136, 158
Photosynthesis, 29–32, 153–154
Pollution, *see* Atmospheric pollution
Potential evaporation, 28
Potential evapotranspiration, 29
Potential temperature, 43–45
Power spectrum, *see* Spectrum of turbulence
Precipitation, 48–49
Profile, 4

Q, R

Quadrature spectrum, 77
Radiation
  direct, 11
  diffuse, 11, 134, 156, 169, 190
  long-wave, 3–4, 17–22, 89–91, 134–135, 137, 169–170, 210
  measurement of, 15–16, 21–22, 170
  net, 3–4, 19–22, 145–146, 155–157, 169–170, 190, 196–199
  reflected, 3–4, 11, 14–16, 133–134, 145, 155, 169, 190, 197, 208–210
  short-wave, 3–4, 9–16, 133–134, 137, 145, 155–156, 169–170, 190, 196–199, 208–210
  solar, *see* short-wave
Reflected radiation, *see* Radiation
Reflectivity, 14
Resistance, 151–153
Respiration, 31
Reynolds number, 55
Richardson
  critical flux number, 86, 87

Richardson (*Continued*)
  flux number, 82, 87
  number, 82, 84, 95–96, 138–140, 209
Roughness length, 59, 65, 108–109, 138, 148–149, 161, 175
Run-off, 28, 158

S

Scale of turbulence, 76, 121
Sea breezes, 179–188
Shear flow, *see* Wind
Shearing stress, 56–59, 65, 77–78, 109
Shelter belts, 210, 212–215
Short-wave radiation, *see* Radiation
Similarity theory, 55–56, 62, 71, 82, 87
Slopes, 182, 189–194
Snow, 133–142
Snow temperature, 136
Soil
  heat transfer, *see* Heat transfer
  moisture, 27–29, 37–41, 158
  temperature, 25–27, 157–158, 215
Solar constant, 9
Solar radiation, *see* Radiation
Specific humidity, *see* Humidity
Spectrum
  of radiation, 8, 12, 18
  of turbulence, 70–74, 76, 78–79, 123
Stability of air, 43, 114, 174, *see also* Lapse rate, Richardson number, Stability ratio
Stability ratio, 82
Stationarity, 68
Steady state condition, 3
Stefan-Boltzmann Law, 8
Structure function, 73
Superadiabatic, *see* Lapse rate
Surface temperature, 23–24
  apparent surface radiative temperature, 17, 23–24, 144–145, 170, 174

T

Taylor's theorem, 121–123
Temperature, *see* Air temperature, Snow temperature, Soil temperature, Surface temperature, Water temperature
Temperature gradient, 43

Temperature profiles, *see* Air temperature, Lapse rate
Time constant, 52
Transfer velocity, 152–153
Transitional states, 107–108, 115–117, 127–128
Transitional zones, 107–115, 127–128, 178–188
Transmissivity, 13, *see also* Beer's law
Transpiration, 29–32, 144–145, 158, 164
Turbidity, 197–198
Turbulence, 66–80
measurement of, 79–80

U, V

Urban heat island, 199–204
Valleys, 182, 189–194
Vapor pressure, *see* Humidity

Viscosity, 56–57
Viscous dissipation, 58–59, 69, 70, 86, 89
Viscous sublayer, *see* Laminar sublayer
Von Kármán's constant, 59–61

W, Z

Water table, 27
Water temperature, 168, 171–172, 174
Weather modification, *see* Modification of local weather
Wilting point, 31
Wind, 53–65, 86–87, 99–106, 108–112, 138, 148, 160–161, 174–176, 179–188, 191–193, 204–205, 212–215
measurement of, 65, 104–106
Zero-plane displacement, 59, 65, 138, 148, 149, 161